Theory and Practice of Tunnel Engineering

Edited by Hasan Tosun

Published in London, United Kingdom

IntechOpen

Supporting open minds since 2005

Theory and Practice of Tunnel Engineering
http://dx.doi.org/10.5772/intechopen.93583
Edited by Hasan Tosun

Contributors
Kaveh Dehghanian, Thien Vo-Minh, Shankar Vikram, Dheeraj Kumar, Duvvuri Satya Subrahmanyam, Thomas Marcher, Georg Erharter, Paul Unterlass, Hasan Tosun

Notice
Statements and opinions expressed in the chapters are these of the individual contributors and not necessarily those of the editors or publisher. No responsibility is accepted for the accuracy of information contained in the published chapters. The publisher assumes no responsibility for any damage or injury to persons or property arising out of the use of any materials, instructions, methods or ideas contained in the book.

First published in London, United Kingdom, 2022 by IntechOpen
IntechOpen is the global imprint of INTECHOPEN LIMITED, registered in England and Wales, registration number: 11086078, 5 Princes Gate Court, London, SW7 2QJ, United Kingdom
Printed in Croatia

British Library Cataloguing-in-Publication Data
A catalogue record for this book is available from the British Library

Additional hard and PDF copies can be obtained from orders@intechopen.com

Theory and Practice of Tunnel Engineering
Edited by Hasan Tosun
p. cm.
Print ISBN 978-1-83969-373-1
Online ISBN 978-1-83969-374-8
eBook (PDF) ISBN 978-1-83969-375-5

We are IntechOpen,
the world's leading publisher of
Open Access books
Built by scientists, for scientists

5,800+
Open access books available

143,000+
International authors and editors

180M+
Downloads

156
Countries delivered to

Our authors are among the

Top 1%
most cited scientists

12.2%
Contributors from top 500 universities

CLARIVATE ANALYTICS
BOOK
CITATION
INDEX
INDEXED

WEB OF SCIENCE™

Selection of our books indexed in the Book Citation Index (BKCI)
in Web of Science Core Collection™

Interested in publishing with us?
Contact book.department@intechopen.com

Numbers displayed above are based on latest data collected.
For more information visit www.intechopen.com

Meet the editor

Hasan Tosun is the president of the Turkish Society on Dam Safety and a former researcher (full professor), in the Civil Engineering Department, Eskisehir Osmangazi University, Turkey. He, as a director, has governed the Earthquake Research Center for 12 years. He was the vice-rector at Uşak University, Turkey, and dean of the Engineering Faculty there. He specializes in geotechnical issues for earth and rockfill dams. Up to 1997, he worked at the General Directorate of State Hydraulic Works and supervised the geotechnical studies of large dams constructed in Turkey. He has published more than 320 technical papers published in national and international journals and conference proceedings. Currently, he is the rector of Mudanya University in Bursa, Turkey and has international memberships for CDA in Canada and for USSD and ASSDO in the United States.

Contents

Preface

Tunnel construction is expensive when compared to the construction of other engineering structures. Detailed surveys indicate that the cost of a tunnel support system ranges between 30 and 50 percent of the total project cost, and can sometimes reach upwards of 70 percent. Currently, theoretical studies and experiences obtained from large projects indicate that costs can be reduced by increasing the efficiency of rock load estimation and support design. In other words, the selection of a support system or systems suitable for rock mass conditions encountered during construction processes plays an important role in reducing project costs. Methods based on rock-support interaction introduce rational solutions for economical and safe tunneling because they provide a good combination of design and construction processes.

A tunnel is not only a hole excavated underground. Unlike other underground openings, a tunnel has a long third dimension relative to its two dimensions in the plane perpendicular to its axis. These structures are a very well-defined balance of the excavated material, the selected support system, and the tasks performed during the excavation phase. For this reason, the physical and geological characteristics of the excavated material such as rock mass behavior, stress history, discontinuity orientation, groundwater, and the type of support systems selected are taken into consideration in the design of the related structure. In addition, other factors that occur during the operation phase such as traffic load, internal pressure, and additional groundwater head are also factors that affect the stability of the structure.

This book includes four excellent contributions on the special issues of tunnel engineering. The overall aim is to improve the theory and practice of the construction of underground structures. The book provides an overview of tunneling technology and includes chapters that address analytical and numerical methods for rock load estimation and design support systems and advances in measurement systems for underground structures. The book discusses the empirical, analytical, and numerical methods of tunneling practice worldwide.

I hope that this book provides an opportunity for young engineers and consultants working in the field of tunneling issues to combine theoretical studies and recent practice in their work.

Dr. Hasan Tosun
Professor,
President,
Mudanya University in Bursa,
Bursa, Turkey

Introductory Chapter: Tunnel Engineering – Rock Load Estimation and Support Design Methods

Hasan Tosun

1. Introduction

The design of an underground opening differs from that of other engineering structures constructed on the ground surface. For subsurface structures, loads caused by overburden pressure and surcharge loadings are taken into account, and support system/systems are designed to meet these total loads, while in surface structures such as dams, bridges, and buildings the designer has to transfer loads caused by the relevant structures to the foundation elements beneath ground. In other words, underground structures are constructed within the ground with many unknown parameters. However, surface structures are designed with materials whose properties are known very well and structure dimensions can be controlled according to the bearing capacity of the ground below the relevant structures. Therefore, rock load estimation is a much more important issue in the design of underground structures. Additionally, typical problems can arise due to the type of rock, structural features in the rock mass, and age of rock formation. For example; older rocks with Precambrian and Paleozoic age can result in huge squeezing pressure while the arching effect cannot be formed in young sedimentary rocks.

In the design of underground structures, as well as rock load estimation, the depth of losing zone, the arching effect, and rock-support interaction also emerge as an important design criteria. For this purpose, so many methods based on country and region facts have been suggested. For example, in North American practice, support design with steel ribs is envisaged depending on the defined simple rock classes, while in Central European tunneling, the design of flexible support systems such as rock bolts, shotcrete, and wire mesh was adopted to form a self-carrying zone around the opening depending on physical and mechanical properties of excavated material, discontinuities of rock mass and mechanical characteristics of support system. In the methods that envisage flexible support design, it is aimed to balance the stresses formed around the underground opening by self-bearing the rock load and to achieve more economical solutions. Numerous studies have been conducted on the reliability and economy of these proposed methods [1–15].

This book includes four excellent contributions on the special issues of Tunnel Engineering. The overall aim of the collection is to improve the theory and practice of underground structures. The articles cover chapters on analytical and numerical methods for rock load estimation and design support systems and advance in the measurement system for underground structures.

2. Empirical methods

There are mainly three separate methods, which are empirically used in rock load estimation and support systems for tunnel structures: (1) Conventional analysis, (2) Geomechanics classification, and (3) Q-system. Some researchers introduced so many studies on the review and comparison of empirical methods [16–23].

2.1 Conventional analysis

The method was originally developed by Terzaghi [24, 25] for a steel rib support system. In the method, it is assumed that the load magnitude for the analytical analysis depends on the rock height, and is supposedly available that rock mass can be dimensioned as a wedge or inclined block. In the Conventional Method external support system, such as steel ribs are selected depending on the recommendation of Terzaghi's rock load concept (nine ground categories). For the conventional analysis, rock load is determined as a function of the height of the loosened rock. In this case, the height of loosened rock referred to Terzaghi's rock load concept was estimated rock at the range of 4.5(B + Ht) to 0, in which B is width and height of the tunnel, respectively and a special definition, which is irrespective of the value of (B + Ht) is given in the method [24]. The Conventional Analysis does not assume a construction sequence.

The model used for the Conventional analysis is mainly based on a graphical solution. In the model of Conventional Analysis, it is assumed that load is radially transferred to the support system and radial deformation does not occur during the loading stage. By this analysis method, the interaction between rock and support system is considered only in developing passive resistance. According to [2], the effects of the relative stiffness of the rock and support system, and the boundary condition between the support and rock are not included in analytical solutions of the conventional analysis.

2.2 Geomechanics classification

Bieniawski [26] suggests a classification system depending on an index (RMR-Rock Mass Rating) for mainly underground openings. The author completed new studies to increase its reliability and to provide optimization on the support system used in tunnels [27–31]. This classification poses two main sections. In the first section, there are five parameters: (1) strength of intact rock material, (2) rock quality designation, (3) spacing of joints, (4) condition of joints, and (5) groundwater conditions. As a strength criterion, compressive strength is utilized. For weak rocks, the index value on point load can be considered instead of uniaxial compressive strength. The aspect of Rock Quality Designation is considered to evaluate the drill core of rock mass. The term "joint" means all discontinuities of rock masses surrounding the opening.

The first section of Geomechanics Classification takes into account the presence of fillings in joints and wall conditions. It also considers continuity and the separation of joints as well as surface roughness. The method introduces a ratio for defining water pressure in joints for 10 m tunnel length or a qualitative criterion for representing groundwater flow around the opening. Ratings are allocated for all ranges of the related parameter. The summation of all ratings introduces an overall rating that represents the crude RMR value of the rock mass for the selected section of the tunnel. The second section considers the joint orientation impact. The crude RMR value is adjusted for considering the influence of joint orientation. The adjusted value, which was called RMR concept in short term, changes within a wide range (from 20 to 100).

Geomechanics Classification emphasizes the orientation of the structural features to the rock mass while taking no account of the rock stress. It has been found that the Geomechanics Classification has difficulties applying in the extremely weak ground, which results in squeezing, swelling, or flowing conditions. However, it introduces a rating obtained from the detailed geological investigation. An empirical equation (ht = [(100-RMR)/100] x B) has been developed to estimate the rock load acting on the support system, based on the RMR of the Geomechanics Classification System [32].

Geomechanical Classification of Bieniawski suggests temporary support systems depending on the RMR value, tunnel characteristics, and excavation method. For very good rock conditions (the RMR value is between 81 and 100) no support is recommended while locally and systematic bolts with shotcrete are considered for good rock (the RMR value is between 61 and 80) and fair rock (the RMR value is between 41 and 60), respectively. For poor rock (the RMR is between 41 and 60) and very poor rock (the RMR value is between 21 and 40), the method suggests the use of steel ribs with the combination of rock bolts and shotcrete. The Geomechanics Classification suggests construction sequences a full face, top heading-beach, and multiple drifts depending on rock mass classes categorized according to RMR values.

2.3 Q-system

Barton et al. [33] empirically introduced a design tool for underground openings, namely the Q-system, which is a geomechanical aspect based on six separate parameters. These are (1) Rock Quality Designation (RQD), (2) joint set number (Jn), (3) joint roughness number (Jr), (4) joint alteration number (Ja), (5) joint water reduction factor (Jw) and (6) stress reduction factor (SRF). The authors developed the system to optimize support requirements without stability problems [34–38]. Recently, NGI [39] introduced a manual for using the Q-system.

The Q-value has been formulated as being three quotients [(RQD/Jn), (Jr/Ja), and (Jw/SRF)] depending on six separate parameters mentioned above. The quotient (RQD/Jn) is defined as a parameter for the block or particle size within a wide range (200 and 0.5). The quotient (Jr/Ja), which is also another parameter that measures inter-block shear strength, introduces valuable data about the roughness and alteration degree of discontinuities. The last quotient (Jw/SRF) consisting of two stress parameters (Jw- joint water reduction factor and SRF- the stress reduction factor) considers water pressure which adversely affects the shear strength of joints and evaluates the loosening load resulting by unloading case through discontinuities and very weak rock.

The equations on support pressure introduced in the Q-system provide a convenient means for developing classification rules for dynamic as well as static loading of underground excavations. The dynamic stresses resulting from the passage of seismic waves may presumably exceed the static stresses by some unknown factors [33, 35]. Q-system does not include the joint orientation as a separate parameter. However, the properties of the most unfavorable joint sets are considered in the assessment of the joint alteration number and the joint roughness.

3. Rational methods based on rock-structure interaction

3.1 Convergence-confinement method (CCM)

The CCM, as a rational method based on rock-support interaction, was first suggested by Ladanyi [40] and then developed by Hoek and Brown [41]. A technical committee approves its recommendation on the CCM [42]. Valuable studies have been realized on the convergence behavior of tunnels by various researchers [43–48].

It considers rock mass behavior to be a tendency to close the excavation. The excavating of the tunnel changes equilibrium conditions in rock mass as well disturbs original stresses. The unloading caused by excavation results in displacements throughout the rock mass. The support system is installed while a change in the original stress occurs and displacements develop. In other words, the temporary support system resists displacements in the surrounding rock during the excavation process. In fact, stresses redistribution and displacement development are controlled by rock-support interaction. This phenomenon is the fundamental principle of the CCM which recognizes the behavior of rock mass during processes of rock excavation and support installation. The convergence behavior of rock mass is represented by a curve that correlates pressure with displacement. For constructing this curve, the strength criterion of rock mass such as primary stress condition, elastic moduli, uniaxial compressive strength, etc. is needed. The ground curve with support characteristic curve provides an excellent design tool to illustrate the geomechanical problems of the project.

In the Convergence-Confinement method, rock load is defined as a function of primary stress conditions, not depending on the height of loosed rock directly. The stress-deformation curve of the surrounding is drawn for estimating the limit pressure to be supported. The CCM introduces an analysis that widely utilizes different support systems. It is available to select all kinds of support systems. However, the selection of supports is based on the ground classes as given in the New Austrian Tunneling Method.

3.2 The new Australian Tunneling method (NATM)

The NATM is also a rational method based on ground-structure interaction as the CCM. It contains design and construction concepts with contractual improvements and poses different items on technical and operational processes. Numerous researchers studied in the NATM to clarify some items on support systems [49–52].

The NATM depends on the principle to reduce support requirements by ground resistance mobilization to optimum case without resulting in any instability. The NATM generally recommends two support systems (outer and inner arch). The outer one has a function as a shell zone having more flexibility to provide stability to the surrounding rock (protective support). The supports suggested for this arch are mainly shotcrete combined with a reinforcement mesh and rock bolts. For the unfavorable ground conditions, the flexible support system mentioned above can be combined with light steel sets. The inner arch generally consists of concrete lining which should be installed after providing equilibrium conditions for outer arch. However, the concrete lining is not installed as a permanent support system prior to the outer arch has reached equilibrium. Rabcewicz and Golser [1] state that this application increases the factor safety if it is needed.

The ground classes empirically relate the geological conditions with the excavation procedure and initial support. The classes used in the preliminary design describe the excavation procedure and the support system in detail and include the quantitative information for designing and construction, although the geological conditions described are qualitative.

The behavior of protective support and surrounding rock during the stress redistribution caused by excavation is controlled by a sophisticated measuring system. Observation of ground and support system provides valuable data for well stabilization of tunnel and optimization on support cost in the NATM [3].

The NATM suggests the utilization of technically advanced support and excavation systems to mobilize the ground resistance to its optimum extent, to redistribute the stresses from the heavily stressed to the less stressed zones, and to improve the material properties of the ground. The flexible supports applied for a relatively short time after excavation accomplish the optimum mobilization of ground resistance, and also provide the redistribution of stresses by a flexible cylinder action as described by Peck [53]. The improvement of ground material is achieved by bolts and shotcrete. The reinforcing action of bolts increases the ground shearing resistance.

The total support capacity is the summation of three components including the resistance of lining and rock bolts, and the resistance of rock arch. The total support capacity should exceed the limit support pressure, which was obtained from the ground-support analysis. Otherwise, the structure will not be appropriately stable and safe. A good example for the NATM is Sanliurfa Tunnels, which consists of two tunnels, each having 26 km long and 7.62 m internal diameter (**Figure 1**).

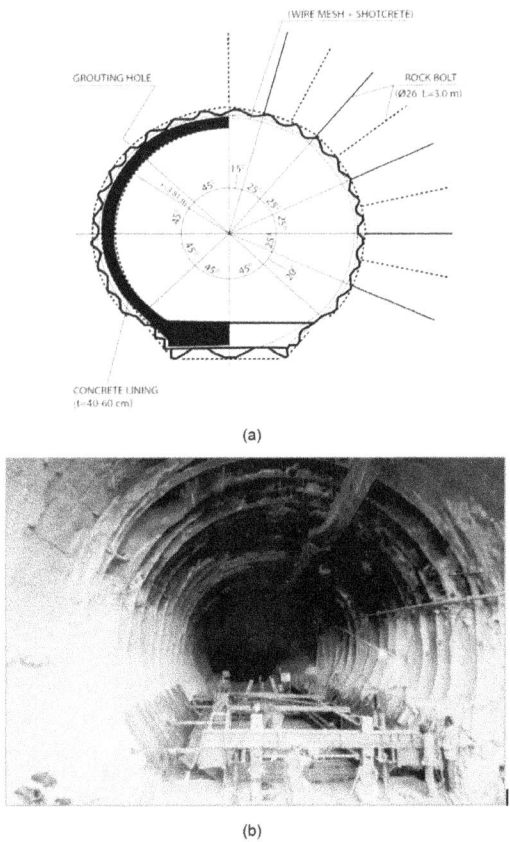

(a)

(b)

Figure 1.
The Sanliurfa tunnel: (a) cross-section of tunnel and (b) a general view from inlet portion of the tunnel.

4. Numerical methods

There are so many numerical approximations for modeling and designing underground structures. The Finite Element Method (FEM), as a most sophisticated numerical analysis, is widely used in analyzing stresses and deformations around an underground opening. In FEM analysis, the structure and the surrounding rock masses are restricted by appropriate boundary conditions and divided into discrete elements, which are triangles and quadrangles connected to each other only at nodes or points of knots. An underground opening can design by the FEM analysis. However, an existing structure can be evaluated as post-failure analysis. It is possible to use it in a wide variety of analyses for sequential construction, control, monitoring, and instrumentation. Recently it is regarded as a common tool for modeling laboratory testing. Numerous studies have been realized on the use of the finite element method for tunneling [54–60]. Examples of the deformation analyses and design of external support system by FEM is given in **Figures 2** and **3**, respectively.

Mathematically it is a numerical technique used for solving differential equations. Stresses and strains for defined elements within the model can be determined

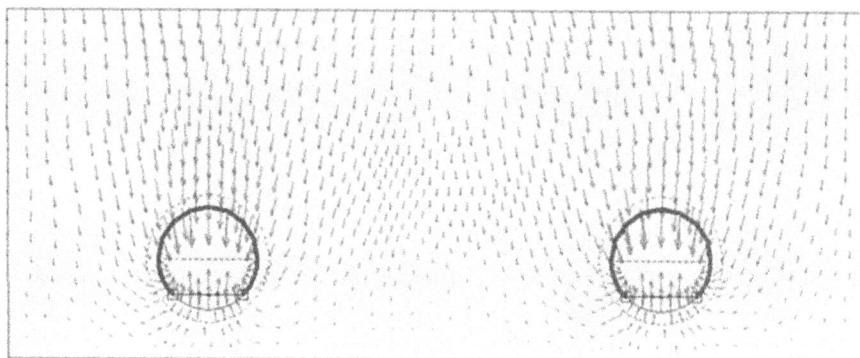

Figure 2.
The deformation (vertical) analyses for the construction second tunnel (unloading) of double tube system of Sanliurfa tunnel by FEM [9, 13–15].

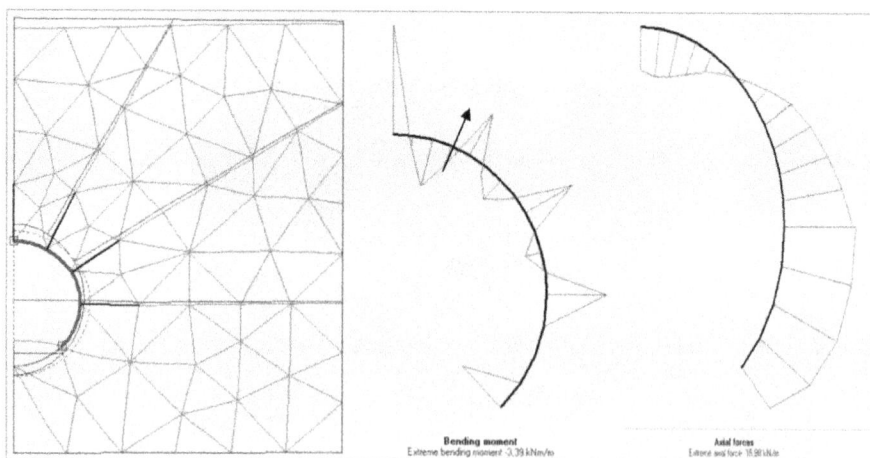

Figure 3.
Design of the external support system (shotcrete+wire mesh+rock bolt) of Sanliurfa tunnel by FEM [9, 13–15].

by the constitutive equations of stress and strain. Physically it is defined as a method for determining element stiffness. In the FEM analyses, a number of alternatives of loading and geometries can be evaluated by two or three-dimensional models. Especially three-dimensional models of underground openings an effectively used for analyzing sequential excavation and support installation. However, the researcher state that the FEM should not be used alone for designing an underground opening [9, 61]. It assists project engineers and consultants in having rational decisions. It poses advantages defining on complex geometry and non-linear nature of geological features as well as providing simplicity for inhomogeneous and discontinuous material [62, 63].

Author details

Hasan Tosun
Mudanya University in Bursa, Bursa, Turkey

*Address all correspondence to: hasan.tosun@mudanya.edu.tr

References

[1] Rabcewicz L, Golser J. Principal of dimensioning the supporting system for the new Austrian Tunneling method. Water Power. 1973:88-93

[2] Agbabian Assoc. Improved Design Procedures for Underground Structural Support Systems in Rock. First Technical Progress Report, R-7814-4803, El Segundo, CA. 1979

[3] Einstein HH, Schwartz CW, Steiner W, Baligh MM, Levitt RE. Improved Design for Tunnel Supports- Analysis Method and Ground Structure Behavior. US Department of Transportation, Report no. DOT/RSPA/ DPB-50/79/10.1980

[4] Tosun H. Parameters considered for tunnel design and geotechnical approaches. In: Proceedings Book. National Seminar on Geotechnics, State Hydraulic Works. Gümüldür-İzmir; 1984. pp. 45-56

[5] Tosun H. Comparison of the tunnel design methods and their application to Urfa aqueduct tunnel. In: MS Thesis in the METU. Ankara; 1985. p. 141

[6] Tosun H. Finite element method for tunnel design. In: Proceedings Book, National Seminar on Geotechnics, General Directorate of State Hydraulic Works. Dragos-İstanbul; 1986. pp. 33-56

[7] Doyuran V, Tosun H. Comparison of tunnel design methods and their application to Urfa tunnel. Middle East Technical University Journal of Pure and Applied Sciences. 1986;19(1):23-61

[8] Doyuran V, Tosun H. Shield Tunneling for soft grounds. In: Proceedings Book of National Seminar on Geotechnics, General Directorate of State Hydraulic Works. Dragos-İstanbul; 1987. pp. 9-40

[9] Tosun H. Use of finite element method on tunnel design. In: Seminar on Use of Software for Hydraulic Engineering. General Directorate of State Hydraulic Works, Sanlıurfa, Proceedings Book. 1991. pp. 1-28

[10] Tosun H. Comparison on design methods of temporary supports using Sanlıurfa tunnels. In: Proceeding of the 2nd Int. Conference on Hydropower. Lillehammer, Norway; 1992. pp. 201-208

[11] Turfan M, Tosun H. Impacts of Geomechanical characteristic of rock masses on support system for pressurized tunnels. In: National Conference on Development of Water and Earth Resources, Proceedings Book. Vol. 2. 1994. pp. 719-729

[12] Tosun H. Design principles for pressurized tunnels and shafts-case studies. Turkish Structure World. 2001;67:23-28

[13] Tosun H. Geotechnical experience obtained from the pressurized tunnel and shafts in Turkey. In: Int. Conference on Tunneling and Underground Space Use, Proceedings Book. Istanbul; 2002. pp. 145-155

[14] Tosun H. Re-analysis of Sanliurfa tunnels by finite element method. In: Proceedings Book, AITES-ITA world tunnel congress- Underground space use: analysis of the past and lessons for the future. Istanbul, Turkey; 2005. pp. 1085-1090

[15] Tosun H. Re-analysis of Sanlıurfa tunnels. International Water Power & Dam Construction. 2007;2007:16-20

[16] Aksoy CO. Review of rock mass rating classification: Historical developments, applications and restrictions. Journal of Mining Science. 2008;44(1):51-63

[17] Rahmani N, Nikbakhtan B, Ahangari K, Apel D. Comparison of

empirical and numerical methods in tunnel stability analysis. International Journal of Mining, Reclamation and Environment. 2012;**26**(3):261-270

[18] Dehkordi MS, Shahriar K, Maarefvand P, Nik MG. Application of the strain energy to estimate the rock load in non-squeezing ground condition. Archives of Mining Sciences. 2011;**56**(3):551-566

[19] Franza A, Marshall AM. Empirical and semi-analytical methods for evaluating tunnelling-induced ground movements in sands. Tunnelling and Underground Space Technology. 2019;**88**:47-62

[20] Khan B, Jamil SM, Jafri TH, Akhtar K. Effects of different empirical tunnel design approaches on rock mass behaviour during tunnel widening. Heliyon. 2019;**5**:e02944

[21] Soufi A, Bahi L, Ouadif L, Kissai JE. Correlation between rock mass rating, Q-system and rock mass index based on field data. In: MATEC Web of Conferences. Vol. 149. 2018. p. 02030

[22] Farhadian H, Nikvar-Hassani A. Development of a new empirical method for tunnel squeezing classification (TSC). Quarterly Journal of Engineering Geology and Hydrogeology. 2020;**53**:655-660

[23] Maleki Z, Farhadian H, Nikvar-Hassani A. Geological hazards in tunnelling: The example of Gelas water conveyance tunnel, Iran. Quarterly Journal of Engineering Geology and Hydrogeology. 2020;**54**(1):qjegh2019-114

[24] Terzaghi K. Introduction to tunnel geology. In: Proctor RV, White TL, editors. Rock Tunnelling with Steel Supports. Youngstown, OH: Commercial Shearing & Stamping Co.; 1946. pp. 17-99

[25] Proctor RV, White TL. Rock Tunneling with Steel Support. Ohio: Commercial Shearing and Stamping Co.; 1968

[26] Bieniawski ZT. Engineering classification of jointed rock masses. The Transactions of the South African Institution of Civil Engineers. 1973;**15**:335-344

[27] Bieniawski ZT. Rock mass classification in rock engineering. In: Bieniawski ZT, editor. Symposium on Exploration for Rock Engineering. Balkema: Rotterdam; 1976. pp. 97-106

[28] Bieniawski ZT. Determining rock mass deformability: Experience from case histories. International Journal of Rock Mechanics and Mining Science and Geomechanics Abstracts. 1978;**15**:237-247

[29] Bieniawski ZT. The geomechanics classification in rock engineering applications. In: Proceedings of the 4th International Congress on Rock Mechanics. Montreux; Vol. 2. 1979. pp. 41-48

[30] Bieniawski ZT. Engineering Rock Mass Classifications: A Complete Manual for Engineers and Geologists in Mining, Civil, and Petroleum Engineering. Vol. xii. New York: Wiley; 1989. p. 251

[31] Bieniawski ZT. Classification of rock masses for engineering: The RMR system and future trends. In: Hudson JA, editor. Comprehensive Rock Engineering, Volume 3. Oxford; New York: Pergamon Press; 1993. pp. 553-573

[32] Singh B, Goel RK. Rock Mass Classification: A Practical Approach in Civil Engineering. Elsevier; 1999

[33] Barton N, Lien R, Lunde J. Engineering classification of rock masses for the design of tunnel support. Rock Mechanics. 1974;**6**:189-239

[34] Barton N. Rock Mass Classification and Tunnel Reinforcement Selection Using the Q-System, Rock Classification for Engineering Purpose. Vol. 984. Philadelphia: ASTM Special Technical Publication; 1988. pp. 59-88

[35] Barton N. The influence of joint properties in modeling jointed rock masses. In: Proceedings of the 8th International Congress on Rock Mechanics. Tokyo, Japan; 1995. pp. 1023-1032

[36] Barton N. Some new Q-value correlations to assist in site characterization and tunnel design. International Journal of Rock Mechanics and Mining Sciences. 2002;**39**:185-216

[37] Barton N, Bandis SC. Review of predictive capability of JRC-JCS model in engineering practice. In: Barton NR, Stephansson O, editors. International Symposium on Rock Joints. Rotterdam: Balkema; 1990. pp. 603-610

[38] Barton N, Løset F, Lien R, Lunde J. Application of the Q-system in design decisions. In: Bergman M, editor. Subsurface space. Vol. 2. New York: Pergamon; 1980. pp. 553-561

[39] NGI. Using the Q-System- Rock Mass Classification and Support Design. Oslo: Norwegian Geotechnical Institute; 2015. p. 54

[40] Ladanyi B. Use of the log-term strength concept in the determination of ground pressure on tunnel lining. In: Proc. 3rd Cong. On Rock Mech., ISRM, v.11, Part B. Denver, Coloroda; 1974

[41] Hoek E, Brown ET. Underground Excavation in Rock: The Institution of Mining and Metallurgy. London, England; 1980

[42] AFTES. In: Panet M, editor. Recommendation on Convergence-Confinement Method. 2001. p. 11

[43] Sinha RS. Underground Structures—Design and Instrumentation. Oxford: Elsevier Science; 1989

[44] Mahdevari S, Torabi SR. Prediction of tunnel convergence using artificial neural networks. Tunnelling and Underground Space Technology. 2012;**28**:218-228

[45] Oreste P. The convergence-confinement method: Roles and limits in modern Geomech. Tunnel design. American Journal of Applied Sciences. 2009;**6**(4):757-771

[46] Alonso E, Alejeno LR, Manin GF, Basance FG. Influence of post-peak properties in the application of the convergence-confinement method for designing underground excavations. In: 5th International Conference and Exhibition on Mass Mining. Luleå, Sweden; 2008

[47] Sandrone F, Labiouse V. Analysis of the evolution of road tunnels equilibrium conditions with a convergence–confinement approach. Rock Mechanics and Rock Engineering. 2010;**43**:201-218

[48] Massinas S. Designing a Tunnel. In: Sakellariou M, editor. Tunnel Engineering. Intechopen; 2019. DOI: 10.5772/intechopen.90182

[49] Rabcewicz L. The New Austrian Tunnelling Method, Part one, Water Power. 1964. pp. 453-457, Part two, Water Power. 1964. pp. 511-515

[50] Muller L. The reasons for unsuccessful applications of the new Austrian tunnelling method, tunnelling under difficult conditions. In: Proceedings of the International Tunnel Symposium. Tokyo: Pergamon Press; 1978. pp. 67-72

[51] Brown ET. Putting the NATM into perspective. Tunnels and Tunneling. November 1981;**13**(10):13-17

[52] Zhang Y, Zuang X, Lackner R. Stability analysis of shotcrete supported crown of NATM tunnels with discontinuity layout optimization. International Journal for Numerical and Analytical Methods in Geomechanics. 2017;**42**(11):1199-1216

[53] Peck RB. Deep excavation and tunneling in soft ground. State of the art report. In: Proc. 7th Int. Conf. on Soil Mech. and Found. Eng. Mexico; 1969. pp. 225-290

[54] Mroueh H, Shahrour I. A full 3-D finite element analysis of tunneling–adjacent structures interaction. Computers and Geotechnics. 2003;**30**:245-253

[55] Roatesi S. Finite element analysis for the problem of tunnel excavation successive phases and lining mounting. In: Chapter in Intechopen Book of Finite Element Analysis-New Trends and Developments. 2012. pp. 301-332

[56] Tang SB, Tang CA. Numerical studies on tunnel floor heave in swelling ground under humid conditions. International Journal of Rock Mechanics and Mining Sciences. 2012;**55**:139-150

[57] Elarabi H, Mustafa A. Comparison of numerical and analytical methods of analysis of tunnels. In: Eighth International Symposium on Geotechnical Aspects of Underground Construction in Soft Ground. IS-Seoul; 2014. pp. 1-8

[58] Zhang ZX, Liu C, Huang X, Kwok CY, Teng L. Three-dimensional finite-element analysis on ground responses during twin-tunnel construction using the URUP method. Tunnelling and Underground Space Technology. 2016;**58**:133-146

[59] Akram MS, Ahmed L, Ullah MF, Rehman F, Ali M. Numerical verification of empirically designed support for a headrace tunnel. Civil Engineering Journal. 2018;**4**(11):2575-2587

[60] Zaid M, Mishra S, Rao KS. Finite element analysis of static loading on urban tunnels. In: Latha Gali M, Raghuveer Rao P, editors. Geotechnical Characterization and Modelling. Lecture Notes in Civil Engineering. Vol. 85. Singapore: Springer; 2020

[61] Kripakov NP. Finite Element Method of Design. Denver, Colorado: U.S. Department of Interior, Bureau of Mines; 1983. p. 17

[62] Swoboda G. Finite element analysis of the new Austrian tunnelling method (NATM). In: Proceedings of the 3rd International Conference on Numerical Methods in Geomechanics. Vol. 2. Aachen; 1979. pp. 581-586

[63] Swoboda G, Marence M, Mader I. Finite element modelling of tunnel excavation. International Journal for Engineering Modelling. 1994;**6**:51-63

Stability Analysis of Circular Tunnels in Cohesive-Frictional Soil Using the Node-Based Smoothed Finite Element Method (NS-FEM)

Thien Vo-Minh

Abstract

In this chapter, the stability of a circular tunnel and dual circular tunnels in cohesive-frictional soils subjected to surcharge loading is investigated by using the node-based smoothed finite element method (NS-FEM). In the NS-FEM, the smoothing strain is calculated over smoothing domains associated with the elements' nodes. The soil is assumed as a uniform Mohr-Coulomb material, and it obeys an associated flow rule. By using the second-order cone programming (SOCP) for solving the optimization problems, the ultimate load and failure mechanisms of the circular tunnel are considered. This chapter discusses the influence of the soil weight $\gamma D/c$, the tunnel diameter ratio to its depth H/D, the vertical and horizontal spacing ratio (L/D, S/D) of two tunnels and soil internal friction angle ϕ on the stability numbers σ_s/c are calculated. The stability numbers obtained from the present approach are compared with the available literature for tunnels.

Keywords: circular tunnel, limit analysis, NS-FEM, SOCP, stability

1. Introduction

In recent years, underground systems have become essential for the rapid development of many big cities. Underground infrastructures as an underground railway and gas pipeline have become increasingly popular in many metropolises to meet public demand. During tunnels' construction, the overburden depth needs to be investigated carefully and plays an important role in constructing process and may reduce construction costs. Therefore, engineers need a practical approach to determine more precise the collapse load and failure mechanism in the circular tunnels' preliminary design stage.

The first studied on the stability of a circular tunnel was performed at Cambridge University in the 1970s. Atkinson and Pott [1], Atkinson and Cairncross [2] investigated a series of centrifuge model tests of tunnels in dry sand and Mohr-Coulomb material subjected to surcharge loading. Cairncross [3] and Seneviratne [4] conducted a series of centrifuge model tests to determine the deformation around a circular tunnel in stiff clay and soft clay. Mair [5], Chambon and Corte [6] also conducted some centrifugal model tests to estimate circular tunnels' stability in soft clay and sandy soil. Recently, Kirsch [7] and Idinger et al. [8] performed a small-scale tunnel model in a geotechnical centrifuge to investigate shallow tunnel face stability dry sand.

Some decades ago, several researchers have studied the stability of a tunnel in cohesive material using the upper and lower bound theorems, for example, the works of Davis et al. [9], Mühlhaus [10], Leca and Dormieux [11]. Recently, Zhang et al. [12] proposed a new 3D failure mechanism using the upper bound limit analysis theory to determine the tunnel face's limit support pressure.. In engineering practice, based on the 3D finite element method, Tosun [13] investigated the performance of concrete lining to compare with those obtained from observation and measurements during the excavation of rock masses and installing the temporary support system.

In recent decades, the finite element method using the triangular element (FEM-T3) has been rapidly developed to solve important geotechnical problems. Sloan and Assadi [14] first applied a finite element procedure for linear analysis to evaluate a square tunnel's undrained stability in a soil whose shear strength increases linearly with depth. Then, Lyamin and Sloan [15], Lyamin et al. [16] and Yamamoto et al. [17, 18] used finite element limit analysis (FELA) to calculate the failure mechanisms of circular and square tunnels in cohesive-frictional soils. Recently, Yamamoto et al. [19], Xiao et al. [20] proposed an efficient method to calculate the stability numbers and failure mechanisms of dual circular tunnels in cohesive material; however, the nonlinear optimization procedure required large computational efforts.

However, one of the drawbacks of FEM-T3 elements is the volumetric locking phenomenon, which is often occurred in the nearly incompressible materials. To overcome this, Chen et al. [21] proposed a stabilized conforming nodal integration using the strain smoothing technique. Then, Liu and his co-workers [22–25] applied this technique to standard FEM and proposed a class of smoothed finite element method (S-FEM). Typical S-FEM models include the cell-based S-FEM model (CS-FEM) [26], node-based S-FEM model (NS-FEM) [27, 28], and edge-based S-FEM model (ES-FEM) [29–32]. Several papers demonstrate that the NS-FEM performs well in heat transfer analysis [33, 34], fracture analysis [35], acoustic problems [36, 37], axisymmetric shell structures [38], static and dynamic analysis [39–41]. Recently, T. Vo-Minh and his co-workers [42–45] applied an upper bound limit analysis using NS-FEM and second-order cone programming (SOCP) to determine the twin circular and dual square tunnels' stability cohesive-frictional soils.

This chapter aims to summarize our research papers [42–45] using the node-based smoothed finite element method (NS-FEM) to estimate the collapse load and failure mechanism of single and two circular tunnels in cohesive-frictional soil subjected to surcharge loading. The numerical results are available for cases with $\phi \leq 30°$, and geotechnical engineers can use them in the preliminary design stage.

2. Upper bound limit analysis for a plane strain with Mohr-coulomb yield criterion using NS-FEM

2.1 A brief on the node-based smoothed finite element method

In the NS-FEM, the problem domain Ω is discretized by N_e triangular elements with totally N_n nodes and N_n smoothing domains $\Omega^{(k)}$ associated with the node k such that $\Omega = \sum_{k=1}^{N_n} \Omega^{(k)}$ and $\Omega^i \cap \Omega^j = \emptyset$, $i \neq j$. The smoothing domain of the node k

in NS-FEM is constructed based on the elements connected to the nodes k, as illustrated in **Figure 1**. The requirement of the smoothing domain is non-overlap and not required to be convex. Therefore, the smoothing domain is created by connecting the mid-edge-points sequentially to the surrounding triangles' centroids.

The smoothed strain associated with the node k in the matrix form can be calculated by

$$\tilde{\varepsilon}_k = \sum_{I \in N^{(k)}} \tilde{\mathbf{B}}_I(\mathbf{x}_k)\mathbf{d}_I \tag{1}$$

where $N^{(k)}$ is a group of nodes associated with smoothing domain $\Omega_k{}^S$, $\mathbf{d}_I$ is the nodal displacement vector and $\tilde{\mathbf{B}}_I(\mathbf{x}_k)$ is the smoothed strain displacement matrix on the smoothing domain $\Omega_k{}^S$ that can be determined as

$$\tilde{\mathbf{B}}_I(\mathbf{x}_k) = \begin{bmatrix} \tilde{b}_{Ix}(\mathbf{x}_k) & 0 \\ 0 & \tilde{b}_{Iy}(\mathbf{x}_k) \\ \tilde{b}_{Iy}(\mathbf{x}_k) & \tilde{b}_{Ix}(\mathbf{x}_k) \end{bmatrix} \tag{2}$$

$$\tilde{b}_{Im}(\mathbf{x}_k) = \frac{1}{A^{(k)}} \int_{\Gamma^{(k)}} n_m^{(k)}(\mathbf{x})N_I(\mathbf{x})d\Gamma, \, (m = x, y) \tag{3}$$

where $A^{(k)} = \int_{\Omega^{(k)}} d\Omega$ is the area of the cell $\Omega_k{}^S$, $n_m^{(k)}(\mathbf{x})$ is a matrix with components of the normal outward vector on the boundary Γ_k, $N_I(\mathbf{x})$ is the FEM shape function for node I.

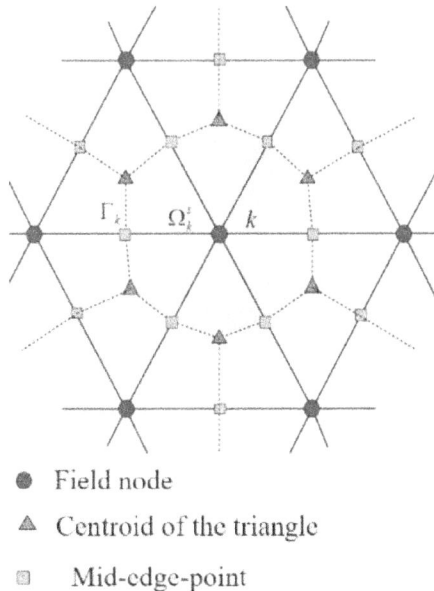

● Field node

▲ Centroid of the triangle

▢ Mid-edge-point

Figure 1.
Triangular elements and smoothing cells associated with nodes.

By using Gauss integration over each sub-boundary Γ_k of $\Omega_k{}^S$, Eq. (3) can be rewritten as

$$\tilde{b}_{Im}(\mathbf{x}_k) = \frac{1}{A^{(k)}} \sum_{j=1}^{n_{eg}} N_I\left(\mathbf{x}_j^{GP}\right) n_{jm}^{(k)} l_j^{(k)}, (m = x, y) \tag{4}$$

where n_{eg} is the total number of the sub-boundary segment of Γ_k, x_j^{GP} is the Gauss point of the sub-boundary segment of Γ_k which has length $l_j^{(k)}$ and outward unit normal $n_{jm}^{(k)}$.

2.2 A brief on the upper bound theorem

Consider a two-dimensional structure made of rigid-perfectly plastic materials with the domain Ω bounded by a continuous boundary $\Gamma_{\dot{u}} \cup \Gamma_t = \Gamma$, $\Gamma_{\dot{u}} \cap \Gamma_t = \emptyset$. The structure subjected to body forces $\mathbf{f}$ and external tractions $\mathbf{g}$ on Γ_t and the boundary $\Gamma_{\dot{u}}$ prescribed by the displacement velocity vector $\dot{\mathbf{u}}$. The strain rates can express as:

$$\dot{\boldsymbol{\varepsilon}} = \left[\dot{\varepsilon}_{xx} \ \dot{\varepsilon}_{yy} \ \dot{\gamma}_{xy}\right]^T = \nabla \dot{\mathbf{u}} \tag{5}$$

The linear form of the external work rate can be calculated by

$$W_{ext}(\dot{\mathbf{u}}) = \int_{\Omega} \mathbf{f} \cdot \dot{\mathbf{u}} d\Omega + \int_{\Gamma_t} \mathbf{g} \cdot \dot{\mathbf{u}} d\Gamma \tag{6}$$

The internal plastic dissipation of the two-dimensional domain Ω can be written as

$$W_{int}(\dot{\boldsymbol{\varepsilon}}) = \int_{\Omega} D(\dot{\boldsymbol{\varepsilon}}) d\Omega = \int_{\Omega} \boldsymbol{\sigma} \cdot \dot{\boldsymbol{\varepsilon}} d\Omega \tag{7}$$

A space of kinematically admissible velocity field is denoted by

$$U = \left\{ \dot{\mathbf{u}} \in \left(H^1(\Omega)\right)^2, \dot{\mathbf{u}} = \bar{\dot{\mathbf{u}}} \ \text{on} \ \Gamma_{\dot{u}} \right\} \tag{8}$$

We define a convex set that contains admissible stress fields

$$\mathbb{S} = \left\{ \boldsymbol{\sigma} \in \sum | \psi(\boldsymbol{\sigma}) \le 0 \right\} \tag{9}$$

where Σ is symmetric stress tensors, $\psi(\boldsymbol{\sigma})$ is the yield function.

The upper bound theorem states that there exists a kinematically admissible displacement field $\dot{\mathbf{u}} \in U$, such that

$$W_{int}(\dot{\boldsymbol{\varepsilon}}) < \alpha^+ W_{ext}(\dot{\mathbf{u}}) + W_{ext}^0(\dot{\mathbf{u}}) \tag{10}$$

where α^+ is the limit load multiplier of the load $\mathbf{g}$, $\mathbf{f}$ and $W_{ext}^0(\dot{\mathbf{u}})$ is the work of additional load $\mathbf{g}_o$, t_o not subject to the multiplier.

Defining $C = \{\dot{\mathbf{u}} \in U | W_{ext}(\dot{\mathbf{u}}) = 1\}$, the upper bound limit analysis becomes the optimization problem to determine the collapse multiplier α^+

$$\alpha^+ = \min \int_\Omega D(\dot{\varepsilon})d\Omega - W^0_{ext}(\dot{u}) \tag{11}$$

$$st \begin{cases} \dot{u} = 0 \ \ \text{on} \ \Gamma_u \\ W_{ext}(\dot{u}) = 1 \end{cases} \tag{12}$$

For plane strain in geotechnical problems, the Mohr-Coulomb yield function can be expressed as

$$\psi(\sigma) = \sqrt{(\sigma_{xx} - \sigma_{yy})^2 + 4\tau^2_{xy}} + (\sigma_{xx} + \sigma_{yy})\sin\phi - 2c\cos\phi \tag{13}$$

For an associated flow rule, the plastic strain rates vector is given by

$$\dot{\varepsilon} = \dot{\mu}\frac{\partial \psi(\sigma)}{\partial \sigma} \tag{14}$$

where $\dot{\mu}$ is the plastic multiplier.

Makrodimopoulos and Martin [46] used the Mohr-Coulomb failure criterion and associated flow rule to determine the power of plastic dissipation in geotechnical problems as follows

$$D(\dot{\varepsilon}) = cA_i t_i \cos\phi \tag{15}$$

where A_i is the area of the element of node i, c is the cohesion, ϕ is the internal friction angle of soil, t_i is a vector of additional variables defined by

$$t_i \geq \sqrt{(\dot{\varepsilon}^i_{xx} - \dot{\varepsilon}^i_{yy})^2 + (\dot{\gamma}^i_{xy})^2} \tag{16}$$

By using the smoothed strains rates $\dot{\tilde{\varepsilon}}_i$ in Eq. (1), the upper bound limit analysis problem for the plane strain using NS-FEM can be discretized in the simple form as follows

$$\alpha^+ = \min \left(\sum_{i=1}^{N_n} cA_i t_i \cos\phi - W^0_{ext}(\dot{u})\right) \tag{17}$$

$$st \begin{cases} \dot{u} = 0 \ \ \text{on} \ \Gamma_u \\ \\ W_{ext}(\dot{u}) = 1 \\ \\ \tilde{\varepsilon}^i_{xx} + \tilde{\varepsilon}^i_{yy} = t_i \sin\phi, i = 1, 2, \ldots \ldots \ldots, N_n \\ \\ t_i \geq \sqrt{(\tilde{\varepsilon}^i_{xx} - \tilde{\varepsilon}^i_{yy})^2 + (\dot{\gamma}^i_{xy})^2}, \ i = 1, 2, \ldots \ldots \ldots, N_n \end{cases} \tag{18}$$

where α^+ is a stability number, and A_i is the area of the smoothing domain of node i, N_n is the total number of nodes in the domain. The fourth constraint in Eq. (18) is a form of quadratic cones. As a result, the conic interior-point optimizer of the academic MOSEK package [47] is used for solving this problem.

3. Numerical examples

In this chapter, GiD [48] software was used to generate triangular elements with reduced element size close to the tunnel's periphery. The domain's size is assumed sufficiently large to eliminate the boundary effects and the plastic zones to be contained entirely within the domain. The computations were performed on a Dell Optiplex 990 (Intel Core™ i5, 1.6GHz CPU, 8GB RAM) in a Window XP environment. The NS-FEM approach has been coded in the Matlab language.

Example 1: Stability of a circular tunnel in cohesive-frictional soil.

Figure 2 shows the analysis model for a plane strain circular tunnel. The tunnel has diameter D and depth H. The soil behavior is modeled as a Mohr-Coulomb material with cohesion c, friction angle ϕ and unit weight γ. The surcharge loading σ_s is applied over the ground surface with smooth and rough interface conditions. **Figure 3** illustrates the half of typical meshes of the upper bound limit analysis problem. The horizontal displacement of nodes is freeing or fixing along the ground surface, respectively, to describe the smooth or rough interface conditions between the loading and the soil. The following equation can calculate the upper bound limit analysis using NS-FEM:

$$\alpha^+ = \frac{\sigma_s}{c} = f\left(\frac{H}{D}, \frac{\gamma D}{c}, \phi\right) \tag{19}$$

Figure 4a shows the plastic dissipation distribution of circular tunnel for shallow tunnel $H/D = 1$, $\phi = 15°$. A failure mechanism originates from the middle part of the tunnel and extends up to the ground surface. The power dissipation for medium and deep tunnels are illustrated in **Figure 4b** and **c**. The collapse surface develops from the bottom of the tunnel and extends up to the ground surface. It is noticeable that the failure mechanisms obtained by this proposed procedure are identical to those derived from rigid blocks and the results of Yamamoto et al. [18]. However, the values of stability number obtained from assuming rigid-block mechanism are greater than those of this proposed numerical procedure.

The stability numbers using NS-FEM for different values of ϕ, $\gamma D/c$ and H/D are listed in **Tables 1** and **2** for smooth and rough interface conditions. The positive

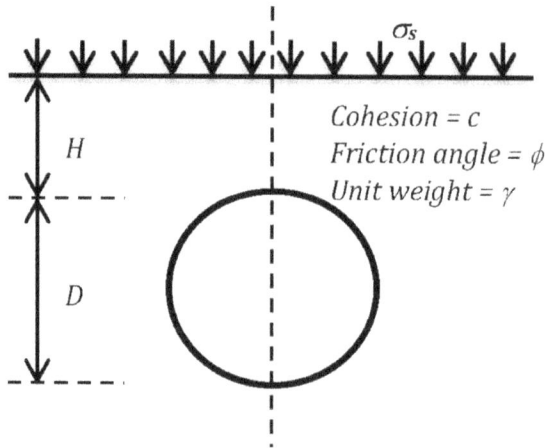

Figure 2.
Model of a circular tunnel subjected to surcharge loading.

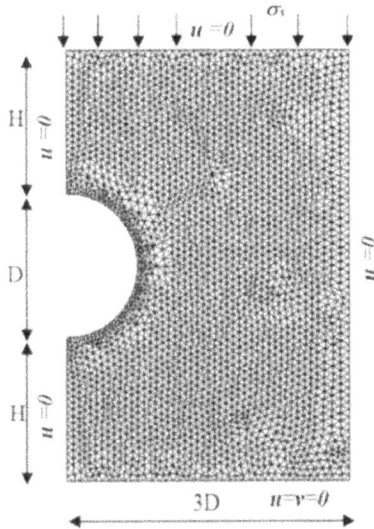

Figure 3.
Typical meshes of the circular tunnel using NS-FEM (H/D = 1).

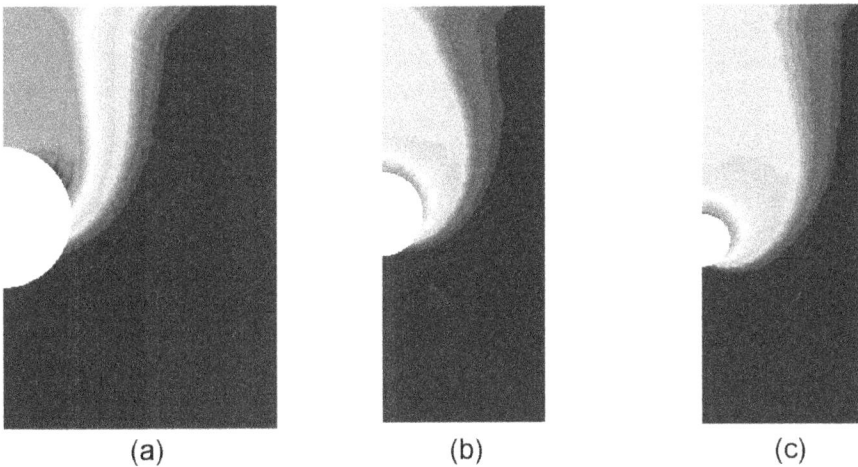

Figure 4.
Plastic dissipation distributions of circular tunnels. (a) H/D = 1, $\phi = 15^o$ $\gamma D/c = 1$. (b) H/D = 2, $\phi = 15^o$ $\gamma D/c = 1$. (c) H/D = 5, $\phi = 15^o$ $\gamma D/c = 1$.

result means that the tunnel collapsed when subjected to compressive stress on the ground surface as this value. The negative value implies that theoretically, only normal tensile stress can be applied to the ground surface to sustain the tunnels' stability. The stability results approximate zero are indicated by "−"; it means that the tunnel failure due to gravity occurs. The stability number results derived from this proposed method agree with the average values of the upper bound and lower bound reported by Yamamoto et al. [18], and illustrated in **Figures 5** and **6**.

Example 2: Stability of dual circular tunnels in cohesive-frictional soil.

Figure 7 illustrates the main geometrical parameters of two circular tunnels under plane strain conditions. The dual circular tunnels with the same diameter D,

ϕ	H/D				$\gamma D/c$			
		0	0.5	1	1.5	2	2.5	3
0	1	2.44	1.85	1.26	0.64	0.02	−0.62	−1.29
	2	3.46	2.33	1.18	0.02	−1.15	−2.35	−3.57
	3	4.13	2.48	0.80	−0.89	−2.60	−4.33	−6.08
	4	4.64	2.47	0.28	−1.92	−4.15	−6.39	−8.65
	5	5.04	2.35	−0.33	−3.07	−5.79	−8.56	−11.31
5	1	2.94	2.30	1.67	1.03	0.38	−0.28	−0.95
	2	4.42	3.18	1.94	0.68	−0.57	−1.84	−3.12
	3	5.47	3.64	1.81	−0.03	−1.88	−3.75	−5.62
	4	6.29	3.88	1.46	−0.96	−3.41	−5.86	−8.34
	5	6.99	3.98	1.00	−2.02	−5.06	−8.12	−11.23
10	1	3.65	2.95	2.26	1.56	0.87	0.15	−0.55
	2	5.88	4.49	3.10	1.70	0.30	−1.12	−2.55
	3	7.64	5.57	3.48	1.37	−0.76	−2.92	−5.13
	4	9.10	6.34	3.55	0.72	−2.16	−5.13	—
	5	10.40	6.95	3.46	−0.12	−3.79	−0.20	—
15	1	4.69	3.92	3.15	2.36	1.59	0.80	0.02
	2	8.31	6.71	5.09	3.45	1.80	0.13	−1.58
	3	11.50	9.06	6.58	4.04	1.44	−1.24	—
	4	14.38	11.08	7.70	4.21	0.59	−3.29	—
	5	17.09	12.92	8.65	4.15	−0.58	−0.21	—
20	1	6.36	5.46	4.57	3.67	2.77	1.85	0.95
	2	12.77	10.81	8.83	6.81	4.76	2.66	0.48
	3	19.25	16.20	13.05	9.79	6.42	2.87	−0.95
	4	25.65	21.47	17.09	12.49	7.63	2.39	—
	5	32.21	26.81	21.20	15.14	8.70	1.48	—
25	1	9.26	8.16	7.09	5.98	4.89	3.76	2.66
	2	21.97	19.42	16.82	14.14	11.39	8.53	5.55
	3	37.34	33.19	28.81	24.22	19.40	14.29	8.81
	4	54.33	48.44	42.14	35.43	28.23	20.43	11.82
	5	73.50	65.64	57.41	48.19	38.44	27.47	15.29
30	1	14.91	13.47	12.09	10.62	9.21	7.72	6.25
	2	44.59	40.88	37.03	33.03	28.64	24.49	19.93
	3	89.47	82.93	75.92	68.42	60.42	51.86	42.65
	4	146.15	136.32	125.60	113.94	101.20	87.35	72.18
	5	218.76	204.65	190.30	172.99	155.12	134.21	112.11
35	1	27.87	25.74	23.73	21.57	19.48	17.27	15.20
	2	114.91	108.36	101.47	94.18	86.50	78.47	69.98
	3	290.04	276.58	261.98	246.33	229.38	211.06	191.18
	4	551.96	530.01	505.28	478.04	447.58	414.36	377.71

ϕ	H/D	$\gamma D/c$						
		0	0.5	1	1.5	2	2.5	3
	5	946.26	910.14	875.14	829.11	783.95	727.41	670.19

Table 1.
Stability numbers σ_s/c of a circular tunnel (smooth interface).

ϕ	H/D	$\gamma D/c$						
		0	0.5	1	1.5	2	2.5	3
0	1	2.51	1.92	1.33	0.72	0.10	−0.55	−1.22
	2	3.53	2.40	1.25	0.08	−1.09	−2.31	−3.53
	3	4.20	2.54	0.87	−0.82	−2.54	−4.27	−6.03
	4	4.70	2.53	0.33	−1.88	−4.11	−6.36	−8.63
	5	5.10	2.41	−0.29	−3.01	−5.75	−8.51	−11.29
5	1	3.03	2.40	1.76	1.11	0.46	−0.20	−0.89
	2	4.51	3.27	2.02	0.76	−0.51	−1.79	−3.07
	3	5.57	3.73	1.89	0.03	−1.83	−3.70	−5.59
	4	6.40	3.98	1.54	−0.89	−3.35	−5.82	−8.31
	5	7.09	4.08	1.07	−1.96	−5.02	−8.10	−11.23
10	1	3.78	3.08	2.38	1.67	0.96	0.24	−0.48
	2	6.04	4.63	3.22	1.80	0.38	−1.06	−2.51
	3	7.80	5.70	3.59	1.46	−0.69	−2.87	−5.10
	4	9.28	6.50	3.68	0.82	−2.09	−5.10	—
	5	10.59	7.11	3.58	−0.03	−3.75	−0.29	—
15	1	4.89	4.10	3.31	2.51	1.71	0.91	0.11
	2	8.58	6.94	5.28	3.61	1.93	0.22	−1.52
	3	11.79	9.30	6.77	4.19	1.55	−1.16	—
	4	14.72	11.37	7.94	4.39	0.70	−3.25	—
	5	17.49	13.26	8.89	4.33	−0.49	−0.25	—
20	1	6.66	5.74	4.82	3.89	2.96	2.03	1.08
	2	13.24	11.23	9.18	7.10	5.00	2.84	0.61
	3	19.80	16.67	13.44	10.11	6.65	3.03	−0.83
	4	26.39	22.10	17.61	12.91	7.94	2.55	—
	5	33.15	27.66	21.89	15.63	9.00	1.64	—
25	1	9.76	8.64	7.52	6.38	5.23	4.08	2.91
	2	22.95	20.30	17.58	14.79	11.93	8.98	5.91
	3	38.59	34.31	29.81	25.08	20.13	14.88	9.26
	4	56.16	50.07	43.56	36.59	29.17	21.12	12.18
	5	76.22	68.28	59.59	50.09	39.84	28.58	15.91
30	1	15.87	14.41	12.92	11.43	9.92	8.36	6.81
	2	46.90	43.02	38.99	34.79	30.43	25.89	21.13

ϕ	H/D	$\gamma D/c$						
		0	0.5	1	1.5	2	2.5	3
	3	93.45	86.72	79.52	71.83	63.67	54.96	45.53
	4	151.97	141.69	130.44	118.30	105.10	90.62	74.69
	5	230.82	216.96	201.32	183.92	164.66	143.29	119.64
35	1	30.01	27.85	25.66	23.42	21.16	18.85	16.49
	2	122.01	115.10	107.88	100.34	92.42	84.10	75.30
	3	309.29	295.79	281.07	265.32	248.11	229.58	209.79
	4	581.11	557.87	532.05	503.36	471.31	436.51	398.22
	5	1038.50	1000.41	964.58	920.20	871.60	817.40	757.44

Table 2.
Stability numbers σ_s/c of a circular tunnel (rough interface).

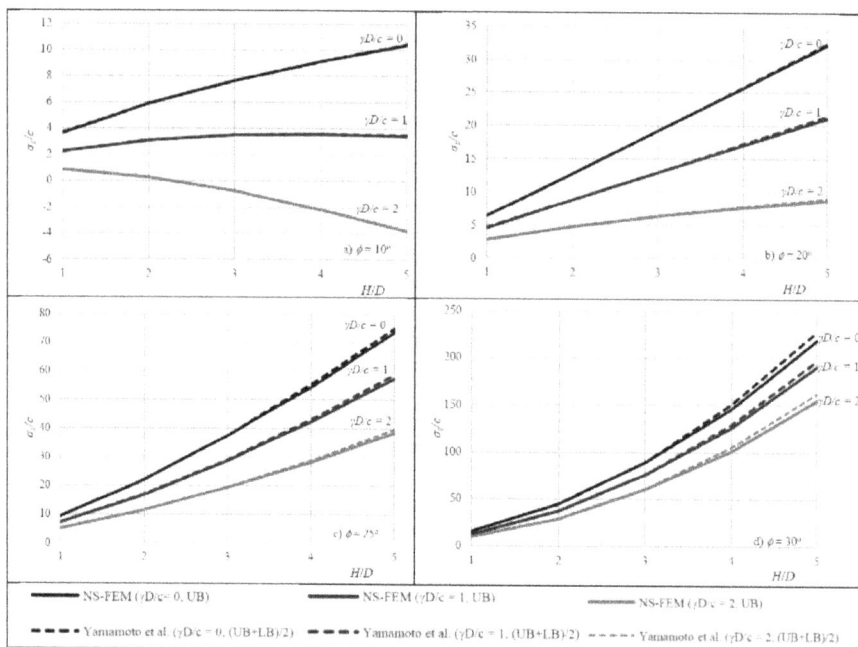

Figure 5.
The variation of stability numbers σ_s/c for different values of H/D (smooth interface).

the cover depth of tunnel H, the vertical L and horizontal S distances between two tunnel centres. Continuous loading σ_s is applied to the ground surface. The soil is assumed to be rigid perfectly plastic and modeled by a Mohr-Coulomb yield criterion with cohesion c, friction angle ϕ, and unit weight γ. The typical mesh of dual circular tunnels is shown in **Figure 8**.

The stability number σ_s/c is defined as a function of ϕ, $\gamma D/c$, S/D, L/D and H/D to investigate two circular tunnels' stability. The following equation can calculate the upper bound limit analysis using NS-FEM:

$$\alpha^+ = \frac{\sigma_s}{c} = f\left(\frac{H}{D}, \frac{L}{D}, \frac{S}{D}, \frac{\gamma D}{c}, \phi\right) \tag{20}$$

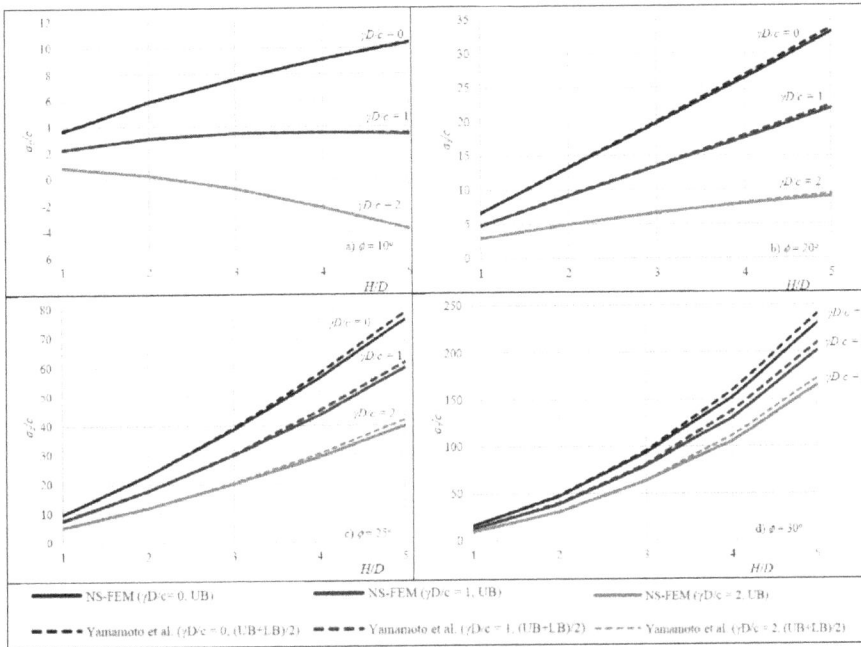

Figure 6.
The variation of stability numbers σ_s/c for different values of H/D *(rough interface).*

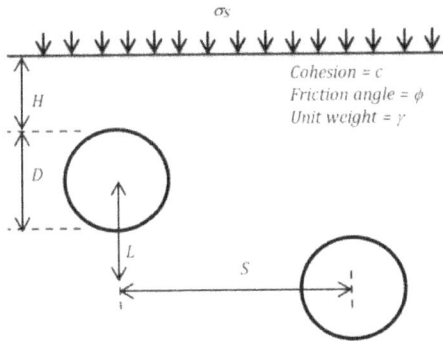

Figure 7.
Model of two circular tunnels subjected to continuous loading.

3.1 Stability of two horizontal circular tunnels ($L/D = 0$, $S/D \neq 0$)

Figure 9a–c show the distribution of power dissipation of shallow tunnel $H/D = 1$, $\phi = 15°$ and $\gamma D/c = 1$ at different values of S/D, i.e., $S/D = 1.5$, 2.0 and 3.5. In **Figure 9a** and **b**, a small slip surface between two circular tunnels enlarges to the top and bottom of tunnels, and a large surface from the middle part of the tunnel extends up to the ground surface. When the distance between two tunnels increases continuously and exceeds a certain value (S_c), i.e., $S \geq S_c = 3.5D$ as shown in **Figure 9c**, the failure mechanism becomes that of two single individual tunnels.

Figure 10 shows the power dissipation of moderate tunnels $H/D = 3$, $\phi = 15°$ and $\gamma D/c = 1$ at different values of S/D, i.e., $S/D = 2.0$, 3.5 and 7.0. In **Figure 10a** and **b**, a

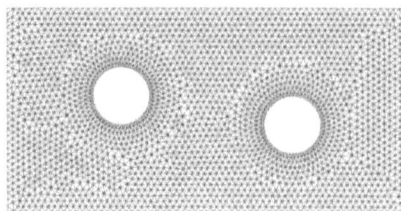

Figure 8.
The typical mesh of two circular tunnels using NS-FEM (H/D = 1, S/D = 3.5, L/D = 0.5).

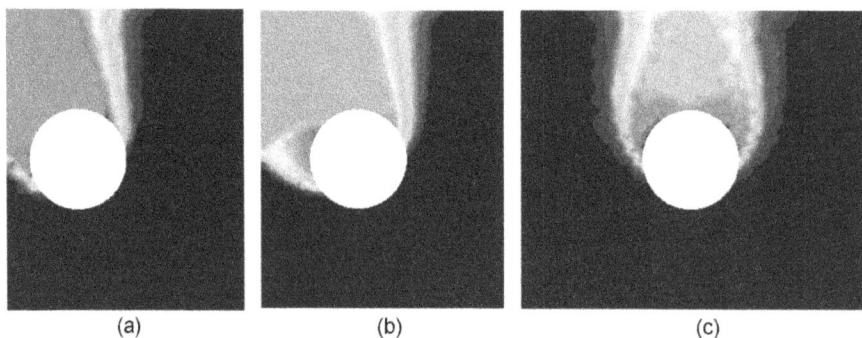

(a)   (b)   (c)

Figure 9.
Power dissipation of dual circular tunnels in the case H/D = 1. (a) $\gamma D/c$ = 1, S/D = 1.5, ϕ = 15°. (b) $\gamma D/c$ = 1, S/D = 2, ϕ = 15°. (c) $\gamma D/c$ = 1, S/D = 3.5, ϕ = 15°.

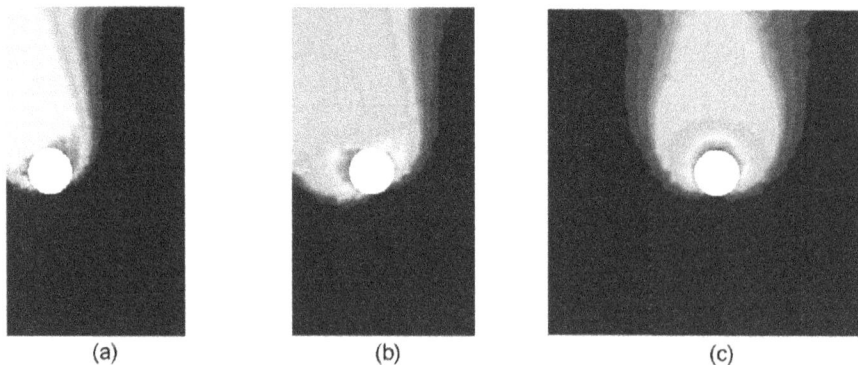

(a)   (b)   (c)

Figure 10.
Power dissipation of dual circular tunnels in the case H/D = 3. (a) $\gamma D/c$ = 1, S/D = 2, ϕ = 15°. (b) $\gamma D/c$ = 1, S/D = 3.5, ϕ = 15°. (c) H/D = 3, $\gamma D/c$ = 1, S/D = 7, ϕ = 15°.

slip failure between two circular tunnels enlarges to the top and bottom of tunnels, and a large surface originates the bottom of the tunnel extends up to the ground surface. When the distance between two tunnels increases continuously and exceeds a critical spacing (S_c), i.e., $S \geq S_c$ = 7D as shown in **Figure 10c**, the failure

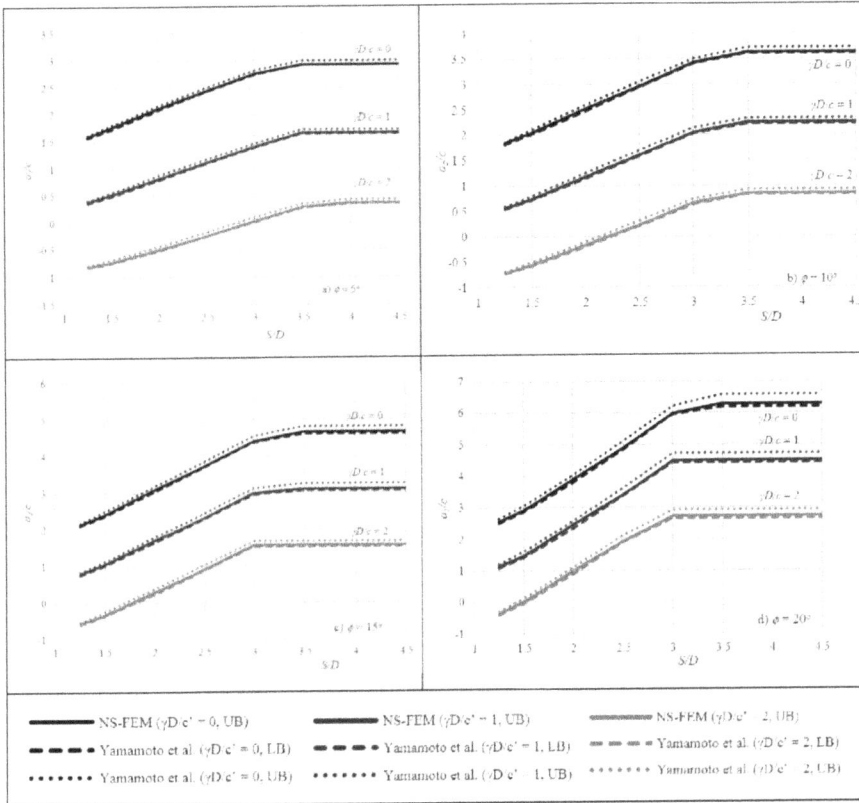

Figure 11.
The variation of stability numbers σ_s/c for H/D = 1 (smooth interface).

mechanism becomes that of two single individual tunnels and no influence on the failure mechanism of each tunnel.

The stability number results using NS-FEM for different values of ϕ, $\gamma D/c$, S/D and H/D are listed in Table A1 of our research paper [43] and shown in **Figures 11** and **12**. The results derived from this proposed method agree well with the average values of the upper bound and lower bound reported by Yamamoto et al. [19]. The stability numbers at the no-interaction points for dual circular tunnels are highlighted in bold. When the spacing between the tunnels exceeds these points, the results obtained from NS-FEM tend to become constant. Therefore, the horizontal distance between two circular tunnels S/D plays an important role in the behavior of the failure mechanism. The increase in stability number is due to the effects of interaction between two circular tunnels.

3.2 Stability of two circular tunnels at different depth ($L/D \neq 0$, $S/D \neq 0$)

Figure 13a shows the typical power dissipation of two circular tunnels for shallow tunnel $H/D = 1$, $\phi = 15°$, $L/D = 0.5$, $S/D = 1.5$. It is noticed that a little slip failure occurs between two tunnels, and a large failure from the middle part of the tunnels extends up to the ground surface. When the distance between two tunnels increases continuously and exceeds a critical spacing (S_c), i.e., $S \geq S_c = 4D$ as shown in **Figure 13b**, only the tunnel near the ground surface failure and no interaction between dual circular tunnels.

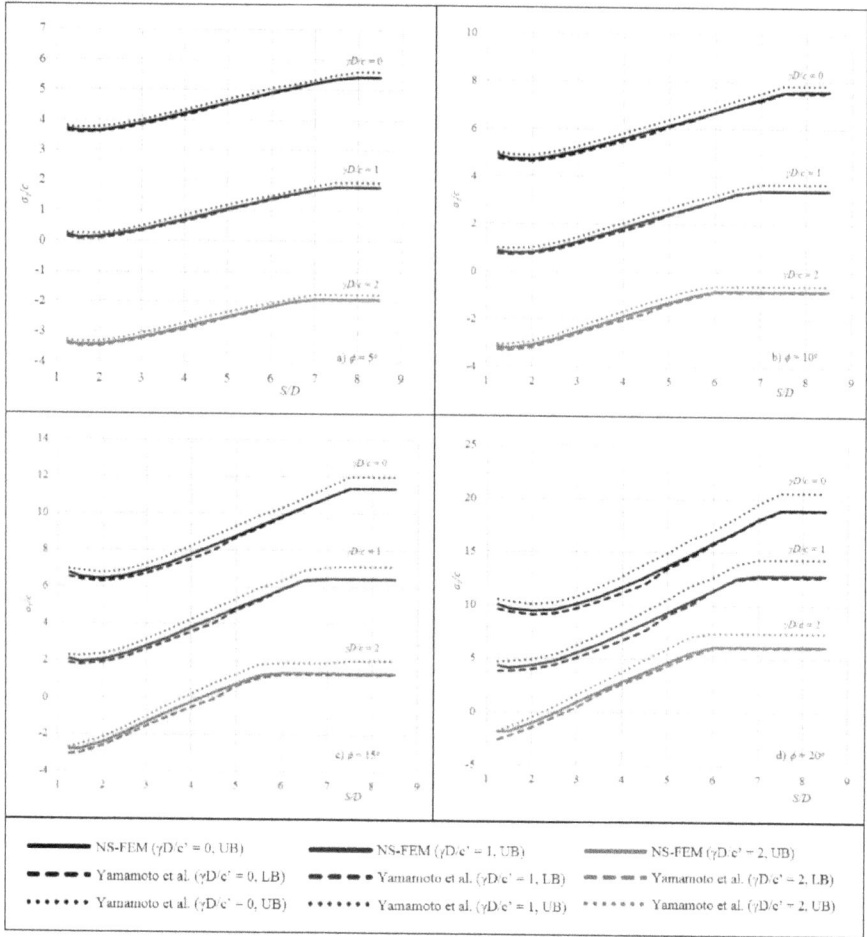

Figure 12.
The variation of stability numbers σ_s/c for H/D = 3 (smooth interface).

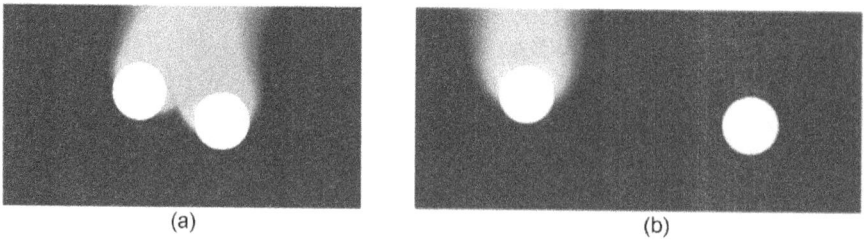

(a)   (b)

Figure 13.
Power dissipation of dual circular tunnels in the case H/D = 1. (a) H/D = 1, L/D = 0.5, S/D = 1.5, γD/c = 1.5, ϕ = 15°. (b) H/D = 1, L/D = 0.5, S/D = 4, γD/c = 1.5, ϕ = 15°.

Figure 14a shows failure mechanism for moderate depth tunnel $H/D = 3, \phi = 15°$, $L/D = 1$, $S/D = 3$. In this figure, a small slip surface between two circular tunnels enlarges to the top and bottom of tunnels, and a large surface from the bottom of the tunnel extends up to the ground surface. When the distance between two

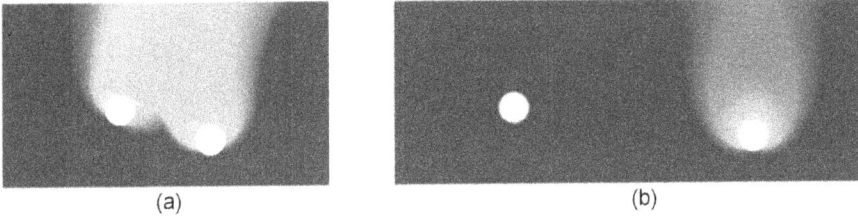

Figure 14.
Power dissipation of dual circular tunnels in the case H/D = 3. (a) H/D = 3, L/D = 1, S/D = 3, γD/c = 1.5,
φ = 15°. (b) H/D = 3, L/D = 1, S/D = 8, γD/c = 1.5, φ = 15°.

tunnels is large enough and exceeds a critical spacing (S_c), i.e., $S \geq S_c = 8D$ as illustrated in **Figure 14b**, only the deeper tunnel failure and no influence on the failure mechanism of each tunnel.

The stability number results using NS-FEM for different values of ϕ, $\gamma D/c$, S/D, L/D and H/D are listed in Tables 2–4 of our research paper [45]. The results derived from this proposed method agree well with the average values of the lower bound and upper bound solution reported by Xiao et al. [20]. The stability numbers at the no-interaction points for dual circular tunnels are highlighted in bold. When the spacing between the tunnels exceeds these points, the results obtained from

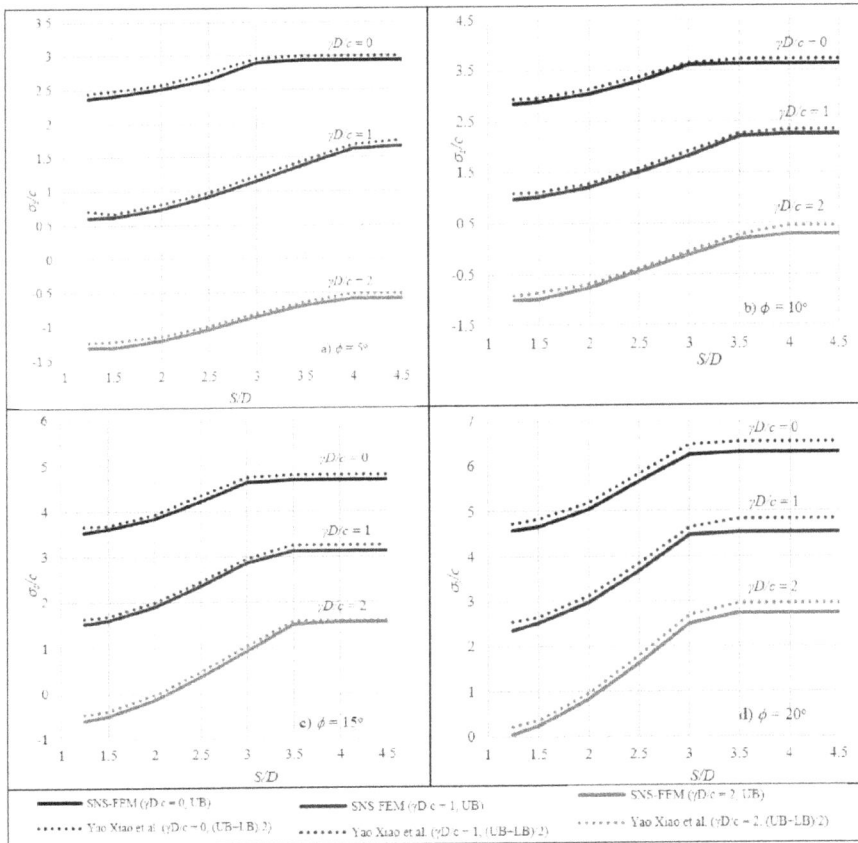

Figure 15.
The variation of stability numbers σ_s/c for H/D = 1, L/D = 1 (smooth interface).

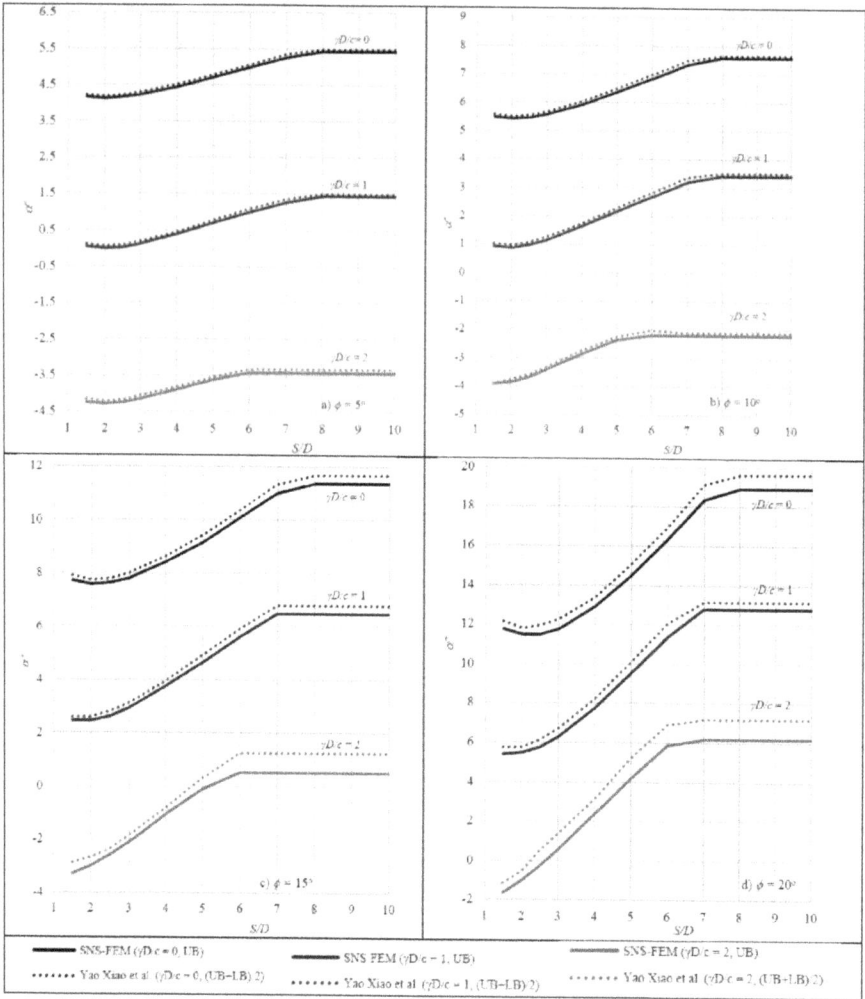

Figure 16.
The variation of stability numbers σ_s/c for H/D = 3, L/D = 1 (smooth interface).

NS-FEM tend to become constant. Therefore, the horizontal distance between two circular tunnels S/D plays an important role in the failure mechanism's behavior. The comparison of stability numbers between NS-FEM and Xiao et al. [20] solution is shown in **Figures 15** and **16**.

3.3 Stability of two vertical circular tunnels ($L/D \neq 0$, $S/D = 0$)

Figure 17 shows the variation of the failure mechanisms in the case $H/D = 1$, $L/D = 1.5$, $\gamma D/c = 1$ with different internal friction angle ϕ. In **Figure 17a**, a large slip surface develops from the middle of the above tunnel, and a slip failure originates from the bottom of the below tunnel extends up to the ground surface. In **Figure 17b–d**, the failure mechanism's width decreases with increasing ϕ and the slip surface only originates from the above tunnel extends up to the ground surface.

The variation of the failure mechanisms in the case $H/D = 1$, $L/D = 3$, $\gamma D/c = 1$ with the different internal friction angle is illustrated in **Figure 18**. When a small

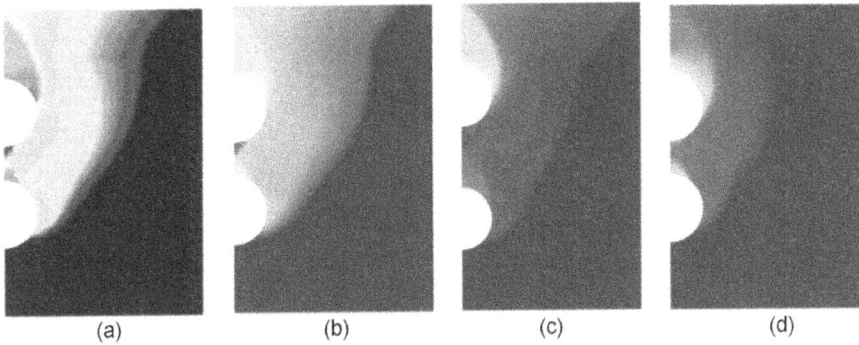

Figure 17.
Power dissipation of dual vertical circular tunnels in the case H/D = 1, L/D = 1.5, γD/c = 1. (a) φ = 0°. (b)
φ = 10°, (c) φ = 20°. (d) φ = 30°.

friction angle $\phi \leq 5°$, a large slip surface develops between two circular tunnels, and a small slip failure originates from the bottom of the below tunnel extends up to the ground surface, shown in **Figure 18a** and **b**. With increasing friction angle $\phi \geq 10°$ shown in **Figure 18c**, the failure mechanism only originates from the above tunnel extends to the ground surface and no failure mechanism on the below tunnel. Therefore, when the vertical distance between two tunnels exceeds a critical spacing (L_c), i.e., $L \geq L_c = 3D$, the failure mechanism occurs with the above shallow tunnel and no effect on the deep tunnel. It means that the below tunnel is more stable when the soil internal friction angle ϕ increase and the slip surface only occurs from the top tunnel.

The stability number results of two vertical using NS-FEM for different values of ϕ, $\gamma D/c$, L/D in the case $H/D = 1$ are summarized in **Table 3**. The stability numbers at the no-interaction points for dual vertical circular tunnels are highlighted in bold. When the spacing between the tunnels exceeds these points, the results obtained from NS-FEM tend to become constant. Therefore, the vertical distance between two circular tunnels L/D plays an important role in the failure mechanism's behavior.

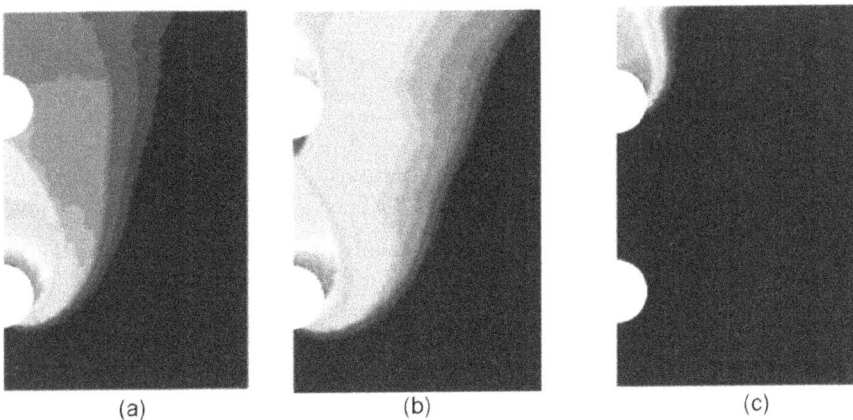

Figure 18.
Power dissipation of dual vertical circular tunnels in the case H/D = 1, L/D = 3, γD/c = 1. (a) φ = 0°. (b)
φ = 5°. (c) φ = 10°, φ = 20°, φ = 30°.

ϕ	L/D				$\gamma D/c$			
		0	0.5	1	1.5	2	2.5	3
0	1.5	2.43	1.77	0.78	−0.34	−1.60	−2.92	−4.31
	2	2.21	1.63	0.48	−1.01	−2.72	−4.59	—
	3	2.43	1.84	0.33	−1.71	−3.93	—	—
	4	2.43	1.84	0.06	−2.28	—	—	—
	5	2.43	1.84	−0.91	—	—	—	—
5	1.5	2.94	2.25	1.28	0.17	−1.06	−2.37	−3.75
	2	2.66	2.07	0.99	−0.42	−2.11	—	−0.21
	3	2.94	2.30	1.22	−0.76	−3.07	—	—
	4	2.94	2.30	1.09	−1.24	—	—	—
	5	2.94	2.30	0.45	−2.98	—	—	—
10	1.5	3.63	2.92	1.93	0.84	−0.41	−1.76	−3.18
	2	3.29	2.64	1.65	0.32	−1.33	—	−0.35
	3	3.63	2.94	2.25	0.54	−1.82	—	—
	4	3.63	2.94	2.25	0.28	−2.57	—	—
	5	3.63	2.94	2.25	−0.91	—	—	—
15	1.5	4.65	3.89	2.85	1.73	0.47	−0.95	−2.47
	2	4.22	3.49	2.58	1.29	−0.26	—	−0.20
	3	4.65	3.89	3.13	2.31	0.20	—	—
	4	4.65	3.89	3.13	2.31	−0.12	—	—
	5	4.65	3.89	3.13	2.31	−1.78	—	—
20	1.5	6.32	5.37	4.26	3.07	1.78	0.30	−1.34
	2	5.70	4.85	3.97	2.71	1.23	−0.59	—
	3	6.32	5.43	4.53	3.64	2.74	0.60	—
	4	6.32	5.43	4.53	3.64	2.74	0.60	—
	5	6.32	5.43	4.53	3.64	2.74	0.60	—
25	1.5	9.12	7.92	6.66	5.32	3.91	2.37	0.58
	2	8.27	7.23	6.19	5.02	3.51	1.82	−0.27
	3	9.18	8.10	7.02	5.94	4.84	3.73	2.61
	4	9.18	8.10	7.02	5.94	4.84	3.73	2.61
	5	9.18	8.10	7.02	5.94	4.84	3.73	2.61
30	1.5	14.29	12.78	11.22	9.60	7.93	6.15	4.22
	2	13.29	11.92	10.53	9.13	7.66	5.85	3.85
	3	14.71	13.32	11.94	10.54	9.08	7.61	6.16
	4	14.71	13.32	11.94	10.54	9.08	7.61	6.16
	5	14.71	13.32	11.94	10.54	9.08	7.61	6.16
35	1.5	25.47	24.14	21.32	19.14	16.89	14.56	12.11
	2	24.77	22.72	20.64	18.55	16.43	14.27	12.05
	3	27.52	25.49	23.43	21.35	19.24	17.08	14.85
	4	27.52	25.49	23.43	21.35	19.24	17.08	14.85

ϕ	L/D	$\gamma D/c$						
		0	0.5	1	1.5	2	2.5	3
	5	27.52	25.49	23.43	21.35	19.24	17.08	14.85

The stability numbers at the no-interaction points for dual vertical circular tunnels are highlighted in bold.

Table 3.
Stability numbers σ_s/c of two vertical circular tunnels (H/D = 1).

4. Conclusions

Based on the upper bound limit analysis using NS-FEM, some concluding remarks can be shown as follow:

1. The stability numbers of a circular tunnel decrease continuously with increasing $\gamma D/c$ and increase with rising parameters H/D and ϕ.

2. A typical failure mechanism of two circular tunnels in cohesive-frictional soils consist of two parts: a small slip surface between the tunnels enlarges to the top and bottom of tunnels, and a large surface from the outside edge of tunnels extends up to the ground surface.

3. The stability results increase with increasing horizontal distance S/D for shallow dual tunnels (H/D = 1). In this case, the stability results increase with increasing horizontal distance S/D until it reaches a critical value S = 3.5D – 4D. The stability number tends to become constant, and this value is exactly equal to that of a single isolated tunnel. For the cases medium and deep tunnels H/D = 3, H/D = 5, the stability number increases until it reaches the approximate values of S = 8D and S = 12D, the stability number becomes constant and arrives at the maximum value.

4. In the case of two circular tunnels at a different depth, the horizontal distance S/D ratio plays an essential role in the behavior of dual circular tunnels' failure mechanisms in cohesive-frictional soils. When the S/D ratio between two tunnels exceeds a certain value, the stability number tends to become constant, while the vertical distance L/D had no significant effect on the stability number. The failure mechanism becomes only a single tunnel and depends on the soil parameters ϕ and $\gamma D/c$.

Acknowledgements

This research was partially supported by the Foundation for Science and Technology at Ho Chi Minh City University of Technology (HUTECH). This support is gratefully acknowledged.

Conflict of interest

The authors declare no conflict of interest.

Author details

Thien Vo-Minh
Faculty of Civil Engineering, Ho Chi Minh City University of Technology
(HUTECH), Vietnam

*Address all correspondence to: vm.thien@hutech.edu.vn

IntechOpen

References

[1] Atkinson J.H, Potts D.M. Stability of a shallow circular tunnel in cohesionless soils. Géotechnique. 1977; 27(2); 203–215.

[2] Atkinson J.H, Cairncross A.M. Collapse of a shallow tunnel in a Mohr-Coulomb material. In: Role of plasticity in soil mechanics, Cambridge; 1973.

[3] Cairncross A.M. Deformation around model tunnels in stiff clay [PhD thesis]. University of Cambridge; 1973.

[4] Seneviratne H.N. Deformations and pore-pressures around model tunnels in soft clay [PhD thesis]. University of Cambridge; 1979.

[5] Mair R.J. Centrifugal modelling of tunnel construction in soft clay [PhD thesis]. University of Cambridge; 1979.

[6] Chambon P, Corté J.F. Shallow tunnels in cohesionless soil: stability of tunnel face. J. Geotech. Eng. 1994; 120 (7); 1148–1165.

[7] Kirsch A. Experimental investigation of the face stability of shallow tunnels in sand. Acta Geotech. 2010; 5; 43–62.

[8] Idiger G, Aklik P, Wu W, Borja R.I. Centrifuge model test on the face stability of shallow tunnel. Acta Geotech. 2011; 6; 105–117.

[9] Davis E.H, Gunn M.J, Mair R.J, Seneviratne H.N. The stability of shallow tunnels and underground openings in cohesive material. Geotechnique. 1980; 30(4); 397–416.

[10] Mühlhaus H.B. Lower bound solutions for circular tunnels in two and three dimensions. Rock Mech Rock Eng. 1985; 18; 37–52.

[11] Leca E, Dormieux L. Upper and lower bound solutions for the face stability of shallow circular tunnels in frictional material. Geotechnique. 1990; 40(4); 581–606.

[12] Zhang C, Han K & Zhang D. Face stability analysis of shallow circular tunnels in cohesive–frictional soils. Tunnelling and Underground Space Technology. 2015; 50; 345–357.

[13] Tosun H. Re-analysis of Sanliurfa tunnel. International Water Power and Dam Construction. 2007; 16–20.

[14] Sloan S.W, Assadi A. Undrained stability of a square tunnel in a soil whose strength increases linearly with depth. Computers and Geotechnics. 1991; 12(4); 321–346.

[15] Lyamin A.V, Sloan S.W. Stability of a plane strain circular tunnel in a cohesive frictional soil. In: Proceedings of the J.R. Booker Memorial Symposium, Sydney, Australia; 2000; p.139–153.

[16] Lyamin A.V, Jack D.L, Sloan S.W. Collapse analysis of square tunnels in cohesive-frictional soils. In: Proceedings of the First Asian-Pacific Congress on Computational Mechanics, Sydney, Australia; 2001; p.405–414.

[17] Yamamoto K, Lyamin A.V, Wilson D.W, Sloan S.W, Abbo A.J. Stability of a single tunnel in cohesive–frictional soil subjected to surcharge loading. Canadian Geotechnical Journal; 2011; 48(12); 1841–1854.

[18] Yamamoto K, Lyamin A.V, Wilson D.W, Sloan S.W, Abbo A.J. Stability of a circular tunnel in cohesive-frictional soil subjected to surcharge loading. Computers and Geotechnics; 2011; 38(4); 504–514.

[19] Yamamoto K, Lyamin A.V, Wilson D.W, Sloan S.W, Abbo A.J. Stability of dual circular tunnels in cohesive–frictional soil subjected to

surcharge loading. Computers and Geotechnics; 2013; 50; 41–54.

[20] Xiao Y, Zhao M, Zhang R, Zhao H, Peng W. Stability of two circular tunnels at different depths in cohesive-frictional soils subjected to surcharge loading. Computers and Geotechnics; 2019; 112; 23–34.

[21] Chen J.S, Wu C.T, Yoon S. A stabilized conforming nodal integration for Galerkin meshfree method. International Journal for Numerical Methods in Engineering; 2001; 50; 435–466.

[22] Liu G.R, Dai K.Y, Nguyen-Thoi T. A smoothed finite element for mechanics problems. Computer and Mechanics; 2007; 39; 859–877.

[23] Liu G.R, Nguyen-Thoi T. Smoothed finite element methods, New York: CRC Press, 2010.

[24] Liu G.R, Nguyen-Xuan H, Nguyen-Thoi T. A theoretical study of S-FEM models: properties, accuracy and convergence rates. International Journal for Numerical Methods in Engineering; 2010; 84; 1222–1256.

[25] Liu G.R, Nguyen-Thoi T, Dai K.Y, Lam K.Y. Theoretical aspects of the smoothed finite element method (SFEM). International Journal for Numerical Methods in Engineering; 2010; 71; 902–930.

[26] Le C.V, Nguyen-Xuan H, Askes H, Bordas S, Rabczuk T, Nguyen-Vinh H. A cell-based smoothed finite element method for kinematic limit analysis. International Journal for Numerical Methods in Engineering; 2010; 83; 1651–1674.

[27] Liu G.R, Nguyen-Thoi T, Nguyen-Xuan H, Lam KY. A node based smoothed finite element method (NS-FEM) for upper bound solution to solid mechanics problems. Computer and Structures; 2009; 87; 14–26.

[28] Nguyen-Xuan H, Rabczuk T, Nguyen-Thoi T, Tran T.N, Nguyen-Thanh N. Computation of limit and shakedown loads using a node-based smoothed finite element method. International Journal for Numerical Methods in Engineering; 2012; 90; 287–310.

[29] Liu G.R., Nguyen-Thoi T, Lam K.Y. An edge-based smoothed finite element method (ES-FEM) for static, free and forced vibration analyses of solids. Journal of Sound and Vibration; 2009; 320; 1100–1130.

[30] Nguyen-Xuan H, Liu G.R. An edge-based finite element method (ES-FEM) with adaptive scaled-bubble functions for plane strain limit analysis. Comput Methods Appl Mech Eng; 2015; 285; 877–905.

[31] Nguyen-Xuan H, Rabczuk T. Adaptive selective ES-FEM limit analysis of cracked plane-strain structures. Frontiers of Structural and Civil Engineering; 2015; 9; 478–490.

[32] Nguyen-Xuan H, Wu C.T, Liu G.R. An adaptive selective ES-FEM for plastic collapse analysis. European Journal of Mechanics A/Solid; 2016; 58; 278–290.

[33] Wu S.C, Liu G.R, Zhang H.O, Xu X, Li Z.R. A node-based smoothed point interpolation method (NS-PIM) for three-dimensional heat transfer problems. International Journal of Thermal Sciences; 2009; 48; 1367–1376.

[34] Cui X.Y, Li Z.C, Feng H, Feng S.Z. Steady and transient heat transfer analysis using a stable node-based smoothed finite element method. International Journal of Thermal Sciences; 2016; 110; 12–25.

[35] Liu G.R, Chen L, Nguyen-Thoi T, Zeng K.Y, Zhang G.Y. A novel singular node-based smoothed finite element method (NS-FEM) for upper bound

solutions of fracture problems. International Journal for Numerical Methods in Engineering; 2010; 83; 1466–1497.

[36] Wang G, Cui X.Y, Liang Z.M, Li G. Y. A coupled smoothed finite element method (S-FEM) for structural-acoustic analysis of shells. Engineering Analysis with Boundary Elements; 2015; 61; 207–217.

[37] Wang G, Cui X.Y, Feng H, Li G.Y. A stable node-based smoothed finite element method for acoustic problems. Computer Methods Applied Mechanics and Engineering; 2015; 297; 348–370.

[38] Cui X.Y, Wang G, Li G.Y. A nodal integration axisymmetric thin shell model using linear interpolation. Applied Mathematical Modelling; 2016; 40; 2720–2742.

[39] Feng H, Cui X.Y, Li G.Y. A stable nodal integration method with strain gradient for static and dynamic analysis of solid mechanics. Engineering Analysis with Boundary Elements; 2016; 62; 78–92.

[40] Wang G, Cui X.Y, Li G.Y. A rotation-free shell formulation using nodal integration for static and dynamic analyses of structures. International Journal for Numerical Methods in Engineering; 2016; 105; 532–560.

[41] G. Wang, X.Y. Cui, G.Y. Li. Temporal stabilization nodal integration method for static and dynamic analyses of Reissner-Mindlin plates. Computers & Structures; 2015; 152; 124–141.

[42] Vo-Minh T, Nguyen-Minh T, Chau-Ngoc A. Upper bound limit analysis of circular tunnel in cohesive-frictional soils using the node-based smoothed finite element method. In: Proceedings of the International Conference on Advances in Computational Mechanics, Phu Quoc Island, Vietnam; 2017; p.123–141.

[43] Thien M. Vo, Tam M. Nguyen, An N. Chau, Hoang C. Nguyen. Stability of twin circular tunnels in cohesive-frictional soil using the node-based smoothed finite element method (NS-FEM). Journal of Vibroengineering; 2017; 19(1); 520–538.

[44] Thien M. Vo, An N. Chau, Tam M. Nguyen, Hoang C. Nguyen. A node-based smoothed finite element method for stability analysis of dual square tunnels in cohesive-frictional soils. Scientia Iranica; 2018; 25(3); 1105–1121.

[45] Vo-Minh T, Nguyen-Son L. A stable node-based smoothed finite element for stability of two circular tunnels at different depths in cohesive-frictional soils. Computers and Geotechnics; 2021; 129; 103865.

[46] Makrodimopoulos A, Martin C.M. Upper bound limit analysis using simplex strain elements and second-order cone programming. International Journal for Numerical and Analytical Methods in Geomechanics; 2006; 31; 835–865.

[47] Mosek, The MOSEK optimization toolbox for MATLAB manual: http://www.mosek.com

[48] GiD 11.0.4, International Center for Numerical Methods in Engineering (CIMNE), Reference manual. http://www.cimne.com

Chapter 3

Analytical Method for Preliminary Seismic Design of Tunnels

Kaveh Dehghanian

Abstract

Buried structures are categorized based on their shape, size and location. These main categories are near surface structures (e.g., pipes and other facilities), large section structures (e.g., tunnels, subways, etc.), and vertical underground structures (e.g., shafts and ducts). Seismic assessments of these structures are important in areas close to severe seismic sources. Seismic design of tunnels requires calculation of the deformation in surrounding geological formations. The seismic hazard on a site is usually expressed as a function of amplitude parameters of free-field motion. Therefore, simplified relations between depth and parameters of ground motion are necessary for preliminary designs. The objective of this chapter is to study and review the main analytical seismic methods which are used to develop a simple relationship between maximum shear strain, maximum shear stress and other seismic parameters.

Keywords: seismic analysis, strain, deformation, free field, analytical methods, tunnel

1. Introduction

A seismic ground motion poses a threat to urban infrastructure as well as human life. Individuals have a limited understanding of underground structures' seismic resistance. Because of smaller deformations under the condition of encompassing rock or soil constraints, it is widely agreed that an underground structure is much more stable than a ground structure. Several communities have emerged in the United States of America to explain seismic behavior of underground opening under severe conditions since the 1990s. Numerous destructive seismic events, such as the Kobe, Chi-Chi, Kocaeli and Wenchuan earthquakes, have occurred since the 1990s, causing genuine harm to tram stations and tunnels, indicating that underground structures are still vulnerable to damage under intense seismic motions. A characteristic example of broad damage due to ground shaking and permanent displacements is the Hanshin earthquake caused liquefaction that contributed to the collapse of numerous underground structures in 1995, counting a tram station in Kobe, Japan, damages to highway tunnels during 1999 Chi- Chi and the collapse of the twin Bolu under construction tunnels, during the 1999 Kocaeli earthquake [1].

Owen and Scholl [2] characterized the deformation sorts of underground structures due to seismic excitation as axial compression/extension; longitudinal bending, ovaling, and racking deformations (**Figure 1**). Shear deformation of tunnels initiated by the vertically propagating shear waves has been broadly investigated by a number of researchers [3, 4], and it has been demonstrated to be the basic mode

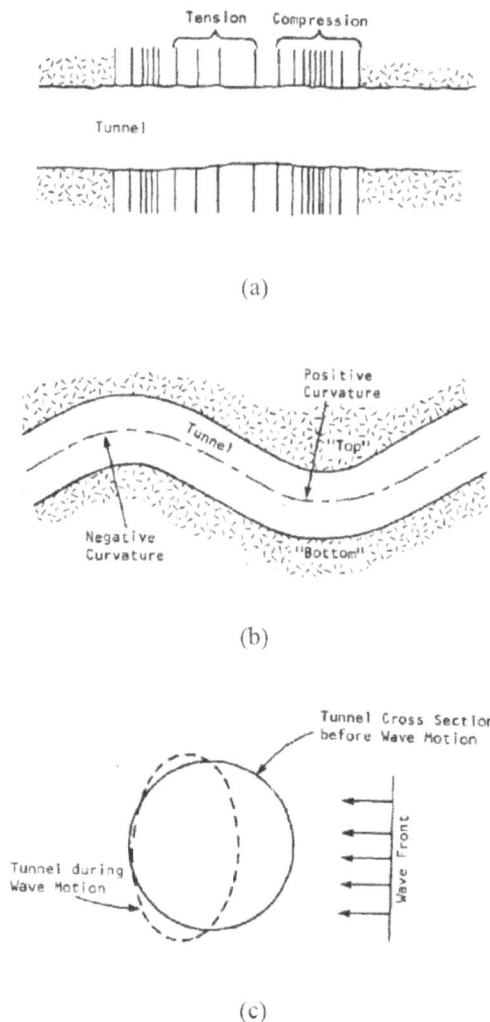

Figure 1.
Types of deformations on tunnels under seismic actions (a) compression extension, (b) longitudinal bending deformation, (c) compression of tunnel section [2].

of deformation for tunnels under seismic loading. Ovaling and racking deformations are related to normal or nearly normal propagation of shear waves with respect to tunnel axes which cause distortion of tunnel cross section. Simplified seismic design approaches for tunnels are often favored by experts. They should be able to assess the general response of a tunnel system that has been subjected to seismic loading. As a result, simpler methods for measuring maximum shear strain (γ_{max}) in the tunnel depth are used [1].

Many researchers proposed analytical solutions to estimate the seismic internal forces of tunnel linings under certain assumptions and conditions, such as elastic response of the soil and tunnel lining, and seismic loading simulation in semi-static construction, among others. Analytical solutions are useful, moderately fast, and easy to use for fundamental seismic design of tunnels, despite the fact that they are formed using relatively strict assumptions and simplifications. As a result, they're commonly used in the early stages of design. With the improvement in technology and computer science, and consequently in numerical analysis of material deformation and stability, several methods are used for analysis of underground structures such as finite element, finite difference and discrete element method. Analyzing of axial and bending deformations can be best performed using 3-D models. In finite difference or finite element models, the tunnel is discretized spatially and the surrounding soil is either discretized or models by springs. Several computer codes perform these type of analysis such as FLAC, ABAQUS and so on [1].

2. Simplified estimation of ground deformations

The seismic design of tunnels is based on two approaches: (1) soil-structure interaction and (2) free field approach. In the first approach, the soil shear strains are affected by the deformation of the nearby underground structures and will conform to the structure strains. A reduction in the total mass of the soil and structure at the soil cavity may have a significant effect on the shear strain. In this case, shear strain of soil in the vicinity of structure will be greater than the free-field approach. In the free-field approach, the interaction between soil and structure is neglected and it is expected that structures accommodate the forced deformations from encompassing ground. These deformations are a function of maximum shear strain [1, 5]. The direct measurement of strains is not possible so it is correlated to other strong-motion parameters such as Peak Ground Velocity (PGV) [6, 7]. Newmark considered one-directional propagation of the harmonic wave in a homogeneous, isotropic, and elastic unbounded medium. According to Newmark, relationship between the maximum particle velocity (V_{max}) and (γ_{max}) is.

$$\gamma_{max} = V_{max}/C \tag{1}$$

Where C is the apparent wave velocity [8].
C cannot be estimated straightforward and is depended on wave type, the angle of incidence, and material property [9]. To calculate this parameter, some formulas are proposed. For instance, O'Rourke and Elhmadi [10] proposed a relation for calculation of longitudinal deformation on buried pipes:

$$C = Vs/sin\emptyset \tag{2}$$

Where $\emptyset$ is angle of the incidence at the ground surface and Vs is the shear wave velocity of the top layer. C is variant at different geological situations [10–12]. Ovaling and racking deformations are correlated with γ_{max} on a vertical plane, so C is close to C_s, which is the incident horizontal shear-wave velocity in geological layers. The consequent structural deformations are basically related to γ_{max} in the imperforated ground as shown in **Figure 2** [13–15].
Wang [13] considering ovaling deformation related C to effective shear modulus, G, and the mass density of the medium, ρ by.

$$C = \sqrt{G/\rho} \tag{3}$$

In the case of replacement of Eq. (3) in Eq. (1) some problems may arise such as the indeterminacy in the definition of deep depth or application of this formula for layered strata. Considering all these issues, they are still adopted by most of the available technical guidelines [6, 7, 12].
St. John and Zahrah [9] developed Newmark's formula and proposed relationships to estimate longitudinal, normal and shear strains in the free field which is depicted in **Table 1**.

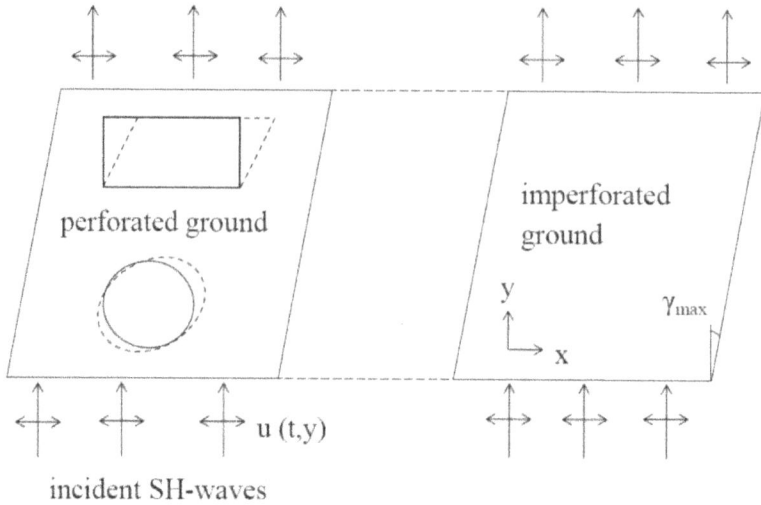

Figure 2.
Ovaling and racking deformation on buried structures [5].

Wave Type		Axial Strain	Shear Strain	Curvature
P-wave		$\varepsilon = \frac{V_P}{C_P}\cos^2\phi$	$\gamma = \frac{V_P}{C_P}\sin\phi\cos\phi$	$\frac{1}{\rho} = \frac{a_P}{C_P^2}\sin\phi\cos^2\phi$
		$\varepsilon_{max} = \frac{V_P}{C_P}$ for $\phi = 0°$	$\gamma_{max} = \frac{V_P}{2C_P}$ for $\phi = 45°$	$\frac{1}{\rho_{max}} = 0.385\frac{a_P}{C_P^2}$ for $\phi = 35.27°$
S-wave		$\varepsilon = \frac{V_S}{C_S}\sin\phi\cos\phi$	$\gamma = \frac{V_S}{C_S}\cos^2\phi$	$\frac{1}{\rho} = \frac{a_S}{C_S^2}\cos^3\phi$
		$\varepsilon_{max} = \frac{V_S}{2C_S}$ for $\phi = 45°$	$\gamma_{max} = \frac{V_S}{C_S}$ for $\phi = 0°$	$\frac{1}{\rho_{max}} = \frac{a_S}{C_S^2}$ for $\phi = 0°$
R-wave	Compressional Component	$\varepsilon = \frac{V_R}{C_R}\cos^2\phi$	$\gamma = \frac{V_R}{C_R}\sin\phi\cos\phi$	$\frac{1}{\rho} = \frac{a_R}{C_R^2}\sin\phi\cos^2\phi$
		$\varepsilon_{max} = \frac{V_R}{C_R}$ for $\phi = 0°$	$\gamma_{max} = \frac{V_R}{2C_R}$ for $\phi = 45°$	$\frac{1}{\rho_{max}} = 0.385\frac{a_R}{C_R^2}$ for $\phi = 35.27°$
	Shear Component		$\gamma = \frac{V_R}{2C_R}\cos\phi$	$\frac{1}{\rho} = \frac{a_R}{C_R^2}\cos^2\phi$
			$\gamma_{max} = \frac{V_R}{C_R}$ for $\phi = 0°$	$\frac{1}{\rho_{max}} = \frac{a_R}{C_R^2}\phi = 0°$

where:

V_P = soil particle velocity caused by P-waves

a_P = soil particle acceleration caused by P-waves

C_P = apparent propagation velocity of P-waves

V_S = soil particle velocity caused by S-waves

a_S = soil particle acceleration caused by S-waves

C_S = apparent propagation velocity of S-waves

V_R = soil particle velocity caused by R-waves

a_R = soil particle acceleration caused by R-waves

C_R = propagation velocity of R-waves

$1/p$ = curvature

Table 1.
Strain and curvature due to body and surface waves [9].

If the shear waves propagate vertically in a uniformly elastic half space, γ_{max} for a specific ground motion is a function of d/Vs, the ratio of depth below free boundary to shear-wave velocity in medium [16]. In layered medium, the equivalent travel-time concept proposed by Imai et al. [17] for estimation of maximum shear-stress (τ_{max}) may be used. Consequently, γ_{max}, can be calculated by dividing τ_{max} by the secant shear modulus of material G_{sec}, representing the average stiffness in a range of shear strain.

$$\gamma_{max} = \tau_{max}/G_{max} \tag{4}$$

For calculation of ovaling deformation, v_{max} is frequently assumed to be equal to the Peak Ground Velocity (PGV) in free field [10, 18]. A reduction coefficient (r_d) is proposed to reduce the ratio of ground motion at tunnel depth to motion at ground surface as it is shown in **Table 2**. This correlation is based on earthquake databases gathered from accelerograms [6, 7].

For tunnels with shallow burial depths, maximum shear stress can be estimated by the product of Peak Ground Acceleration (PGA) in ground surface and overburden pressure [7]. This product is corrected by an empirical depth-reduction factor (r_d) due to the deformability of medium [19]. In this method, maximum shear stress (on a horizontal plane) at depth d is.

$$\tau_{max} = PGA.\rho.d.r_d \tag{5}$$

such that ρ is the density of the shallow geological formation, and d is the depth of interest. Then, maximum can be estimated by Eq. (3).

Penzien [20] also suggested closed-form solutions for seismic analysis of deep rectangular and circular tunnels, with the seismic loading being better replicated as a uniform shear-strain dissemination, τ_{ff}, forced on the soil boundaries of the soil-tunnel system, away from the tunnel. Penzien's solutions, on the other hand, ignore the impact of typical stresses generated during loading along the soil-tunnel interface. They decided that the deformation of the tunnel could be approximated by the deformations of a circular cavity (e.g. through significant consideration of parameter β in **Figure 3**). Huo et al. [21] proposed improved arrangements by considering the genuine deformation example of rectangular-molded cavities and representing both the ordinary and shear stresses at the the soil-tunnel interface.

Analytical solutions usually presume that the soil has a linear elastic behavior and therefore do not take into account the strain-dependent soil shear modulus. Bobet et al. [22] compensated for the reduction in shear modulus by iteratively adjusting the soil shear modulus as a function of shear strain magnitude before shear strain convergence was achieved. The analytical solution was then used to estimate the soil deformation using the compatible shear strain shear modulus [21]. The effect of soil saturation was overlooked in the production of all of the above

Tunnel Depth (m)	Ratio of Ground Motion at Tunnel Depth to Motion at Ground Surface (r_d)
≤ 6	1.0
6 to 15	0.9
15 to 30	0.8
> 30	0.7

Table 2.
Ratios of ground motion at tunnel depth to motion at ground surface [6, 7].

(a)   (b)

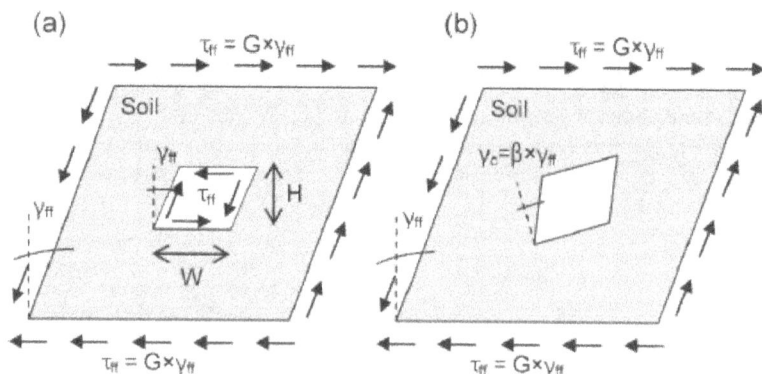

Figure 3.
Deformation of W × H rectangular cavity subjected to a uniform shear strain distribution γ_{ff} (a) with free-field shear stress distribution applied to cavity surface; (b) with free-field shear stress distribution removed from cavity surface [20] (G: soil shear modulus, γ_c: shear distortion of cavity without the application of shear stress distribution around the cavity, $\beta = \gamma_c/\gamma_{ff}$).

closed-form solutions. Bobet [4] suggested circular tunnel solutions in saturated soil, assuming a non-slip interface. Bobet [23] went on to extend the previous solutions to look at the response of rectangular tunnels under no-slip and fully-slip interface conditions, as well as drained and undrained soil conditions. Park et al. [24, 25] looked over the previous solutions and proposed a new approach for considering future sliding along the soil-tunnel interface. The majority of the above-mentioned suggested analytical relationships are for shear S-waves propagating upward in the tunnel's transverse direction. Kouretzis et al. [26–29] proposed a set of relations for compressional P-wave tunnels as well.

The assumptions on which the analytical solutions are based limit their applicability (**Table 3**). Researchers started comparing the results of analytical solutions

Solution	Tunnel lining	Soil type	Saturation conditions	Soil layering	Soil-tunnel interface			Cross-section
	Elastic	Elastic	Dry	Homogeneous	No slip	Frictional Slip	Full Slip	Circular
St.John C.M. and Zahrah T.F [9]	Yes	Yes	Yes	Yes	Yes	No	Yes	Yes
Wang, J.N., [13]	Yes	Yes	Yes	Yes	Yes	No	Yes	Yes
Penzien and Wu [31]	Yes	Yes	Yes	Yes	Yes	No	Yes	Yes
Penzien [20]	Yes	Yes	Yes	Yes	Yes	No	Yes	Yes
Bobet [4]	Yes	Yes	Yes	Yes	Yes	No	Yes	Yes
Hou, et al. [21]	Yes	Yes	Yes	Yes	Yes	No	Yes	No
Park et al. [25]	Yes	Yes	Yes	Yes	Yes	Yes	Yes	Yes
Bobet [32]	Yes	Yes	Yes	Yes	Yes	No	Yes	Yes
Kouretzis [27]	Yes	Yes	Yes	Yes	Yes	No	Yes	Yes
Kouretzis [28]	Yes	Yes	Yes	Yes	No	No	Yes	Yes
Kouretzis [29]	Yes	Yes	Yes	Yes	No	No	Yes	Yes

Table 3.
Summary of assumptions and applicability of analytical solutions for the analysis of tunnels under ground shaking [30].

with the predictions of sophisticated numerical models after the rapid growth of computational power in the last two decades to recognize the shortcomings of these analytical solutions. For example, Kontoe et al. [15] compared four different analytical models (i.e. [13, 20, 23, 24]) and validated them against finite element simulations (FE). Tsinidis et al. [33] compared the results of analytical solutions (i.e. [13, 20, 24]) with numerical predictions for extreme lining flexibilities, i.e. very flexible or very rigid tunnels compared to the surrounding soil. Kontoe et al. [14] and Tsinidis et al. [33] found that the analytical solution of Penzien [20] underestimates the thrust added to the tunnel structure for a slip-free interface, which is consistent with previous findings [34]. As a result, using this solution for a rough soil-lining interface is not recommended.

Since the soil response is often assumed to be linearly elastic, the solutions are usually more reliable only when the soil undergoes minor deformations, such as for very rigid clays and rocks at low shaking levels, with the exception of Bobet et al. [22]. The solutions for the transverse earthquake response are derived in the plane strain condition and therefore cannot be used for complex ground plans. In most cases, the contact interface is limited to two extreme states, full or no slip, while the lining is assumed to be continuous; therefore, a suitable representation of the segmental lining by an equivalent continuous lining is mandatory.

3. Application of random vibration theory in estimation of γ_{max}

Random vibration theory (RVT) relates the statistical properties of the random behavior of a dynamical system to the system properties or those of the random excitation. Therefore, RVT can be used to statistically estimate the random response of a system by representing the ground motion by a power spectral density (PSD) function.

Simplified theoretical conclusions are possible by assuming that ground motion is a stationary (i.e., the statistical properties of the motion are constant in time) Gaussian process. Although earthquake excitations are not stationary, the strong phase of such motions can be assumed to be stationary [35]. In this approach, the excitation is first defined by a PSD. The response PSD is either expressed theoretically or calculated using transfer functions. Then the statistical properties of the response are estimated using its PSD.

A well- known example of the use of RVT for the development of theoretical solutions is the Complete Quadratic Combination (CQC) method, which is useful for estimating peak displacements or forces within a structure [36]. CQC is also used for analyzing the nonstationary random responses of complex structures that are in an inhomogeneous stochastic field [37]. The analysis of the seismic response of linear multicolumn structural systems can be formulated by RVT, which takes into account the multicolumn input [38]. The steady-state filtered white noise model proposed by Kanai and Tajimi [39, 40] provides a well-known PSD in the field of earthquake engineering. White noise is a stationary random process that has a mean of zero and a constant spectral density for all frequencies. In the Kanai-Tajimi spectral model, the rock acceleration is assumed to be white noise and the overlying ground deposits are simulated by a linear one-degree-of-freedom system. Modified Kanai-Tajimi models are also proposed in the literature [41]. Therefore, RVT can be used to generate simple theoretical solutions. On the other hand, these simple solutions are limited to linear systems.

The theorems of random oscillation can be used to derive theoretical relationships between the parameters of dynamic response and ground motion. The theoretical analysis of the random response can be simplified by two assumptions. The first is that the excitation is statistically stationary in a broad sense. The second assumption is that the probability distribution of the excitation is Gaussian, so that each linear operation on this random process produces a different Gaussian process [42]. Although the properties of transient seismic motions obviously contradict these assumptions, the simplification can lead to reasonable theoretical functions that reflect the characteristic properties of dynamical systems. The applications concerning the combination of maximum modal displacements in structural dynamics [36, 43] and transfer functions for kinematic soil-structure interaction [44, 45] are well-known examples.

4. Conclusion

Analytical methods are implemented for analyzing underground structures by a numerous researchers. Though these methods have some shortcomings because of simplifying the design conditions, they provide a good approximation for preliminary analysis of such structures. Analytical methods are divided into two main categories: (a) soil-structure interaction and (b) free-field methods. In this chapter, free-field method, which ignores interaction between structure and encompassing soil, is being studied and its development has been discussed. For the practitioner, the simplified techniques are useful tools for preliminary studies. They make it simple to identify the variables that influence the severity of the prejudices, providing insight into the structure's actions. Furthermore, the simplified approach and its solutions are invaluable in better understanding the relationship between dynamic loads, viscoelastic foundations, and tunnel structures, defining the most important parameters for the problem, and providing preliminary estimates or even a design. They also have the advantage of being able to conduct sensitivity analyses with little effort. The simplified approach may not be able to capture the responses and damage in structural specifics, components, or positions of possible failure due to the simplified assumptions for the tunnel layout and soil-tunnel interaction.

Conflict of interest

No potential conflict of interest was reported by the author.

Author details

Kaveh Dehghanian
Istanbul Aydin University, Istanbul, Turkey

*Address all correspondence to: kavehdehghanian@aydin.edu.tr

IntechOpen

References

[1] Hashash, YMA, Hook, JJ, Schmidt, B and Yao JIC., Seismic Design and Analysis of Underground Structures. Tunneling and Underground Space Technology, Vol.16, No.4, pp. 247–293, 2001.

[2] Owen G.N., Scholl R.E., Earthquake Engineering of Large Underground Structures. Report No. FHWA/RD 80/195, Federal Highway Administration, Washington, DC, 1981; 38-73.

[3] Amorosi A., Boldini D, Falcone G., Numerical Prediction of Tunnel Performance during Centrifuge Dynamic Tests. Acta Geotechnica, Vol.9, No.4, pp.581-596, 2014.

[4] Bobet, A., 2003. Effect of pore water pressure on tunnel support during static and seismic loading. Tunn. Undergr. Space Technol. 18 (4), 377–393.

[5] Niyazi E., 2010. Analysis Of Seismic Behavior Of Underground Structures: A Case Study On Bolu Tunnels, A Thesis Submitted to the Graduate School Of Natural And Applied Sciences of Middle East Technical University, Ankara.

[6] Federal Highway Administration (FHWA)., Seismic Retrofitting Manual for Highway Structures: part 2 – Retaining Structures, Slopes, Tunnels, Culverts and Roadways. Final Report No. FHWA-HRT-05-067. Multidisciplinary Center for Earthquake Engineering Research, University at Buffalo: New York, pp.162-163, 2004.

[7] Federal Highway Administration (FHWA)., Technical Manual for Design and Construction of Road Tunnels- Civil Elements. Publication No. FHWA-NHI-10-034, the United States, pp.1-702, 2009.

[8] Newmark N.M. Problems in Wave Propagation in Soil and Rock: Symposium on Wave Propagation and Dynamic Properties of Earth Materials, August 23-25, Univ. of New Mexico Press, pp.7-26, 1968.

[9] St. John C.M. and Zahrah T.F., Aseismic Design of Underground Structures. Tunneling and Underground Space Technology, Vol. 2, No. 2, pp.165-197, 1987.

[10] O'Rourke, M. and El Hmadi, K., Analysis of Continuous Buried Pipelines for Seismic Wave Effects. Earthquake Engineering and Structural Dynamics, Vol.16, No.6, pp.917-929, 1988.

[11] Paolucci, R. and Pitilakis, K. Seismic Risk Assessment of Underground Structures under Transient Ground Deformations: Earthquake Geotechnical Engineering, Volume 6 of the series Geotechnical, Geological and Earthquake Engineering. pp.433-459, 2007.

[12] CEN, Eurocode 8-Design of Structures for Earthquake Resistance. Part 4: Silos, Tanks and Pipelines. EN 1998-4, Final draft, January 2006, Brussels.

[13] Wang, J.N., Seismic Design of Tunnels, a Simple State-of-the-Art Design Approach. Parsons Brinckerhoff Inc. Publication, New York, pp.53-133, 1993.

[14] Kontoe, S, Zdravkovic, L, Potts, D. M. and Menkiti, C.O., On the Relative Merits of Simple and Advanced Constitutive Models in Dynamic Analysis of Tunnels. Geotechnique, Vol.61, No.10, pp.815-829, 2011.

[15] Kontoe, S, Avgerinos, V. and Potts, D.M., Numerical Validation of Analytical Solutions and Their Use for Equivalent Linear Seismic Analysis of Circular Tunnels. Soil Dynamics and Earthquake Engineering, Vol.66, No.11, pp.206-219, 2014.

[16] Chen, CH. and Hou, PC., Response Spectrum of Ground Shear Strain. 10th World Conference on Earthquake Engineering, Rotterdam, 1992.

[17] Imai T, Tonouchi K. and Kanemori T., the Simple Evaluation Method of Shear Stress Generated by Earthquakes in Soil Ground. Bureau of Practical Geological Investigation: Japan, pp. 39-58, 1981.

[18] Paolucci, R. and Smerzini, C. Earthquake-induced Transient Ground Strains from Dense Seismic Networks. Earthquake Spectra, Vol. 24, No.2, pp.453–470, 2008.

[19] Seed H.B., Idriss I.M., Simplified Procedure for Evaluating Soil Liquefaction Potential. Journal of Soil Mechanics and Foundations Division, ASCE, Vol.107, No.9, pp.1249-1274, 1971.

[20] Penzien, J., 2000. Seismically induced racking of tunnel linings. Earthquake Eng. Struct. Dyn. 29, 683–691.

[21] Hou, J.S., Tseng, D.J., Lee, Y.H., 2007. Monitoring and evaluation after repair and reinforcement of damaged 3-lane Freeway Tunnel located within fault influenced zone. In: Barták, J., Hrdina, I., Romancov, G., Zlámal, J. (Eds.), Underground Space – the 4th Dimension of Metropolises. Taylor and Francis Group, London, pp. 1885–1890.

[22] Bobet, A., Fernandez, G., Huo, H., Ramirez, J., 2008. A practical iterative procedure to estimate seismic-induced deformations of shallow rectangular structures. Can. Geotech. J. 45, 923–938.

[23] Bobet, A., 2010. Drained and undrained response of deep tunnels subjected to far-field shear loading. Tunn. Undergr. Space Technol. 25 (1), 21–31.

[24] Park, D., Sagong, M., Kwak, D.Y., Jeong, C.G., 2009a. Simulation of tunnel

response under spatially varying ground motion. Soil Dyn. Earthquake Eng. 29, 1417–1424.

[25] Park, K.H., Tantayopin, K., Tontavanich, B., Owatsiriwong, A., 2009b. Analytical solution for seismic-induced ovaling of circular tunnel lining under no-slip interface conditions: a revisit. Tunn. Undergr. Space Technol. 24 (2), 231–235.

[26] Kouretzis, G., Sloan, S., Carter, J., 2013. Effect of interface friction on tunnel liner internal forces due to seismic S- and P-wave propagation. Soil Dyn. Earthquake Eng. 46, 41–51.

[27] Kouretzis, G.P., Bouckovalas, G.D., Gantes, C.J., 2006. 3-D shell analysis of cylindrical underground structures under seismic shear (S) wave action. Soil Dyn. Earthquake Eng. 26 (10), 909–921.

[28] Kouretzis, G.P., Bouckovalas, G.D., Karamitros, D.K., 2011. Seismic verification of long cylindrical underground structures considering Rayleigh wave effects. Tunn. Undergr. Space Technol. 26 (6), 789–794.

[29] Kouretzis, G.P., Andrianopoulos, K. I., Sloan, S.W., Carter, J.P., 2014. Analysis of circular tunnels due to seismic P-wave propagation, with emphasis on unreinforced concrete liners. Comput. Geotech. 55, 187–194.

[30] Grigorios Tsinidis, Filomena de Silva, Ioannis Anastasopoulos, et.al. 2020. Seismic behaviour of tunnels: From experiments to analysis, Tunnelling and Underground Space Technology 99(May), 1-20.

[31] Penzien, J., Wu, C., 1998. Stresses in linings of bored tunnels. Earthquake Eng. Struct. Dyn. 27, 283–300.

[32] Bobet, A., 2010. Drained and undrained response of deep tunnels subjected to far-field shear loading.

Tunn. Undergr. Space Technol. 25 (1), 21–31.

[33] Tsinidis, G., Pitilakis, K., Anagnostopoulos, C., 2016c. Circular tunnels in sand: Dynamic response and efficiency of seismic analysis methods at extreme lining flexibilities. Bull. Earthq. Eng. 14 (10), 2903–2929.

[34] Hashash, Y.M.A., Park, D., Yao, J.I. C., 2005. Ovaling deformations of circular tunnels under seismic loading, an update on seismic design and analysis of underground structures. Tunn. Undergr. Space Technol. 20 (5), 435–441.

[35] Der Kiureghian A., A Response Spectrum Method for Random Vibration, Report No. UCB/ EERC-80/ 15, Earthquake Engineering Research Center, Berkeley, California, pp.1-31, 1980.

[36] Wilson E.L., Der Kiureghian A., Bayo E.P., A Replacement for the SRSS Method in Seismic Analysis, Earthquake Engineering and Structural Dynamics, Vol.9, No.2, pp.l87-l92, 1981.

[37] Lin, J., Li, J., Zhang, W., and Williams, F.W., Random seismic responses of multi-suppo0rt structures in evolutionary in homogenous random fields. Earthquake Engineering and Structural Dynamics, Vol.26, No.1, pp.135-145, 1997.

[38] Heredia-Zavoni, E., Vanmarcke, E. H., Seismic random vibration analysis of multi- support structural systems. Journal of Engineering Mechanics, Vol. 120, No.5, pp.1107-1128, 1994.

[39] Kanai, K., Semi-empirical formula for the seismic characteristics of the ground motion. Bulletin of the Earthquake Research Institute, University of Tokyo Vol.35, No.2, pp.308-325, 1957.

[40] Tajimi, H., A statistical method of determining the maximum response of a building structure during an earthquake. Second World Conference on Earthquake Engineering, Vol.2, session 2, pp.781-798, 1960.

[41] Clough, R.W., Penzien, J., Dynamics of structures, McGraw-Hill, New York, 2003.

[42] Yang C.Y., Random Vibration of Structures. John Wiley & Sons: Canada, pp.201-205, 1986.

[43] Amini A., Trifunac M.D., Statistical extension of response spectrum superposition. Soil Dynamics and Earthquake Engineering, Vol.4, No.2, pp. 54-63, 1985.

[44] Luco JE,Wong HL. Response of a rigid foundation to a spatially random ground motion. Earthquake Engineering and Structural Dynamics, Vol.14, No.6, 1986.

[45] Veletsos A.S., Prasad A.M., Wu W. H., Transfer functions for rigid rectangular foundations. Earthquake Engineering and Structural Dynamics, Vol.26, No.1, pp.5- 17, 1997.

Chapter 4

Capabilities and Challenges Using Machine Learning in Tunnelling

Thomas Marcher, Georg Erharter and Paul Unterlass

Abstract

Digitalization changes the design and operational processes in tunnelling. The way of gathering geological data in the field of tunnelling, the methods of rock mass classification as well as the application of tunnel design analyses, tunnel construction processes and tunnel maintenance will be influenced by this digital transformation. The ongoing digitalization in tunnelling through applications like building information modelling and artificial intelligence, addressing a variety of difficult tasks, is moving forward. Increasing overall amounts of data (big data), combined with the ease to access strong computing powers, are leading to a sharp increase in the successful application of data analytics and techniques of artificial intelligence. Artificial Intelligence now arrives also in the fields of geotechnical engineering, tunnelling and engineering geology. The chapter focuses on the potential for machine learning methods – a branch of Artificial Intelligence - in tunnelling. Examples will show that training artificial neural networks in a supervised manner works and yields valuable information. Unsupervised machine learning approaches will be also discussed, where the final classification is not imposed upon the data, but learned from it. Finally, reinforcement learning seems to be trendsetting but not being in use for specific tunnel applications yet.

Keywords: Big Data, TBM tunnelling, NATM, Automatic Classification, Machine Learning

1. Introduction

Digitisation in tunnelling is an ongoing process that draws on developments in Machine learning (ML) (a sub-field of artificial intelligence -AI) or advanced life cycle systems like building information modelling (BIM). While ML techniques have been used in other disciplines for some time, the demand for ML applications in geotechnics and tunnelling is growing more slowly. Many of the publications using ML for problem solving in geotechnical engineering or tunnelling rely on supervised ML; with [1–3] three papers are given that use artificial neural networks (ANN) to classify rock mass behaviour using tunnel boring machine (TBM) operational data.

The main drawback for those applications in geotechnical engineering is the limited availability of sufficient amounts of high quality data. To this day, only a small portion of the theoretically available data is in use during the design and construction process of tunnels (regardless of whether this data is stored for documentation purposes or is obtained as a by-product of construction works). Unfortunately, such data till now is never used to its full extent and a clear methodology for objective and comprehensible data analysis is lacking. This applies specifically to geological and geotechnical applications, where many classifications are inherently semi-quantitative. Especially the bias introduced by man-made categorical classification presents a great challenge [2].

Great potential is therefore seen in unsupervised ML, where the final classification is learned from the data rather than imposed on it. ML techniques can be used to improve the efficiency and self-consistency of daily work in tunnel design and construction [4].

Finally, reinforcement learning (RL), another branch of ML, seems to be in vogue. To our knowledge, this form of ML has not yet been used for specific applications in geotechnical engineering and tunnelling. Basically, RL refers to the process of an agent learning to achieve a specific goal through interaction with its environment.

Two important prerequisites must be explicitly pointed out regarding data source and quality of the data:

- before processing data with ML techniques, the source of the data has to be verified and data preparation/pre-processing has to be performed (raw data must be separated from inaccurate or irrelevant parts of the data set).

- ethical use by all involved parties is imperative to provide the necessary safety required to get the most out of this technology [2].

Digital transformation in underground construction will be achieved through digital data, automation and networks. This transformation will affect both conventional and mechanised tunnelling. This change will influence payment and contract models, as well as software solutions for tunnel construction in general.

2. The future of digitised tunnel design and construction

The future of digitisation in tunnelling lies in a fully digitised project organisation linking different key technologies, e.g.:

- Machine learning (ML),

- Building Information Model (BIM),

- Augmented Reality (AR).

Through using machine learning techniques, it will mainly be possible to: (1) perform fully autonomous support installation, (2) elicit automatic rock classification, (3) update the geological forecasts in front of the tunnel excavation face (prior to arrive with the tunnel excavation), (4) overcome limitations in the definition of constitutive behaviour of soil and rock, explore the applicability of RL to fully automate different construction processes (self-driving TBMs).

The use of BIM will have an enormous impact on the design, construction and operation of tunnel projects. However, current developments in BIM for tunnelling are mostly focused on the basics of BIM: 3D geometries and corresponding data models /semantics. To fully implement the transition from "simple" semantically enriched 3D geometries to full digital twins, involving the above given technologies is imperative as only this allows for the necessary information exchange within the model. Digital transformation is achieved through systematic data collection and automation and will influence both conventional (sequential) and continuous (TBM) tunnel constructions.

During the planning phase, digital data acquisition, data management and 3D modelling techniques will improve the way geological models or rock mechanics prediction models are created for tunnel projects [5]. This change will influence payment and contract models and will require the systematic implementation of software solutions for construction in general.

Finally, AR can be expected to become more widespread throughout the field of tunnelling. It gives a view of the real world where elements and layers are super-imposed by computer generated files such as graphics, sounds, videos, or other digital information. This computer technology offers significant benefits through simulation and visualisation in the construction industry, e.g. by allowing the user to directly immerse him–/herself in specific information of the environment. Users can interact with both actual and virtual objects and monitor construction progress by contrasting the planned (target) state with the actual state of the project [6]. The users of AR may experience the enhanced world while digital information, including virtual models and contextual information, is presented and augmented with the real world [7]. In areas such as engineering, entertainment, aerospace, medicine, military, and automotive industry, AR technologies have been used as a frontline technology to meet visualisation difficulties in their specific domain [8]. These technologies still have considerable need for research. Their full potential is not fully reached yet [9].

3. Machine learning

3.1 Overview

Machine Learning is a sub-field of the research for AI and deep learning is itself a sub-discipline of ML (**Figure 1**). Where AI research in general focuses on under-standing and synthesising intelligence, deep learning is a specific field that uses multilayer computational frameworks such as artificial neural networks (ANNs) to learn from data. The tremendous advances of ML in the past years (e.g. object

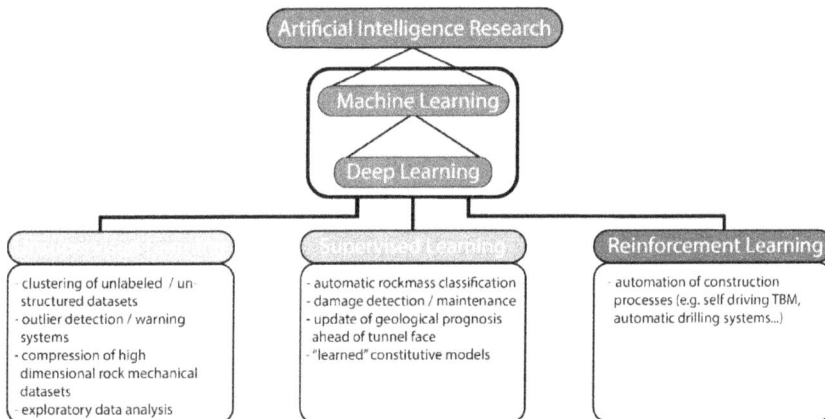

Figure 1.
The fields of artificial intelligence, machine learning and deep learning in a topical context to each other as well as possible applications of the three sub-branches — supervised, unsupervised and reinforcement learning — of ML in tunnelling (modified after [4]).

detection, speech recognition etc.) are mostly based on this technology as it provides a high performing way of establishing input – output connections. However, downsides of deep learning are for example its "data hungry" nature (the impressive functionalities of deep learning are only possible through tremendous datasets) and the "black box" characteristics of the algorithms themselves, where the learned reasoning and logics are still poorly understood. ML itself is comprised of three main branches — supervised learning, unsupervised learning, reinforcement learning — which are described below.

3.2 Supervised learning

Supervised learning is the most widely applied type of ML with common applications being regression and classification tasks. To train supervised learning algorithms labelled datasets are required. Therefore, the input and the output values have to be known before the algorithm is trained (for further information see [10]). If such sufficient datasets are provided, state of the art algorithms can achieve great performance and are theoretically able to learn almost every possible relationship. The dependence on datasets with predefined input and output is however also a downside of supervised learning, as many real world datasets are inherently unlabelled and labelling them is either impossible or very expensive (see next chapter for more information).

The input can usually be imagined as a vector quantity [11] consisting of multiple features. These features are consigned to the learning algorithms together with the corresponding output and during training the algorithm learns to establish an input – output function. For evaluation of the training progress, the whole dataset is divided into several parts where one is used for model training, one for model validation during the training and in some cases a third independent dataset is split off for the sake of testing after the training process is finished. This partitioning of the dataset is necessary as supervised learning algorithms have a tendency of overfitting the data, they are trained on which ultimately leads to a bad generalisation performance if the algorithm is confronted with unseen data.

During training, the model learns a function that is able to map the given input to the corresponding output [11] (**Figure 2**). Supervised learning has already been applied for various geotechnical applications and in tunnelling (e.g. [1–3] natural hazards (e.g. [12]) and constitutive modelling (e.g. [13]).

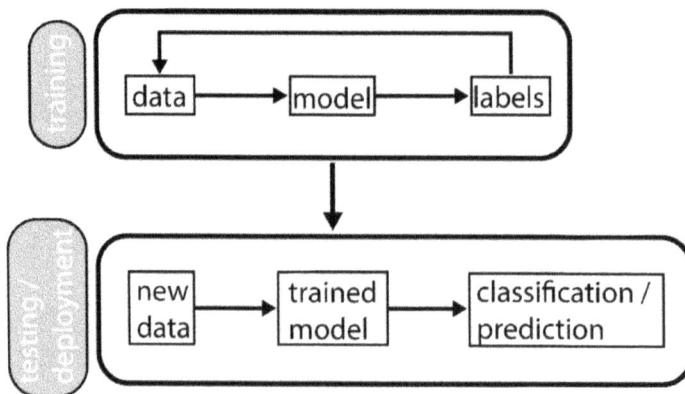

Figure 2.
Basic principle of supervised learning (modified after [4]).

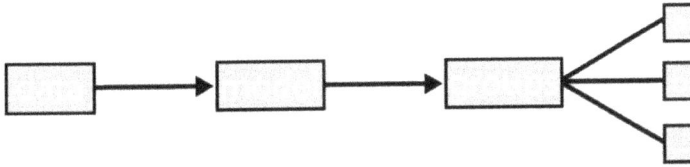

Figure 3.
Basic principle of unsupervised learning.

3.3 Unsupervised learning

Unsupervised learning is a sub-category of machine learning for which the algorithms receive only inputs but no labelled data. The aim of unsupervised ML is for the machine to build representations of the data [14] that in the end helps the operator to gather new information about the dataset. In the course of unsupervised ML, almost all steps can be viewed as learning a probabilistic model of the data [15] (**Figure 3**). The main methods of unsupervised learning and possible geotechnical applications are outlier detection (e.g. for monitoring works), clustering (e.g. to identify structure within data [16] or applying K-Means clustering to recognise rock mass types within TBM operational data) and dimensionality reduction to visualise high dimensional space in a more comprehensible way [14] (e.g. for improving the performance of geophysical log data classification).

3.4 Reinforcement learning

While in supervised and unsupervised learning the data is the main focus and algorithms either learn from or about it, reinforcement learning (RL) is about algorithms that improve their performance from interaction with the environment [17]. Algorithms/models are often called "agent" in this case and can be thought of as players of entities that can take certain action to influence the overall state of their surroundings. The environment on the other hand is the agents' battleground which changes as a response to their actions and provides feedback to them by sending an updated state back to the agent and a reward signal that allows the agent to assess its own performance (**Figure 4**). The agent initially begins with performing random actions and over time starts to learn a "policy" for completing a task by

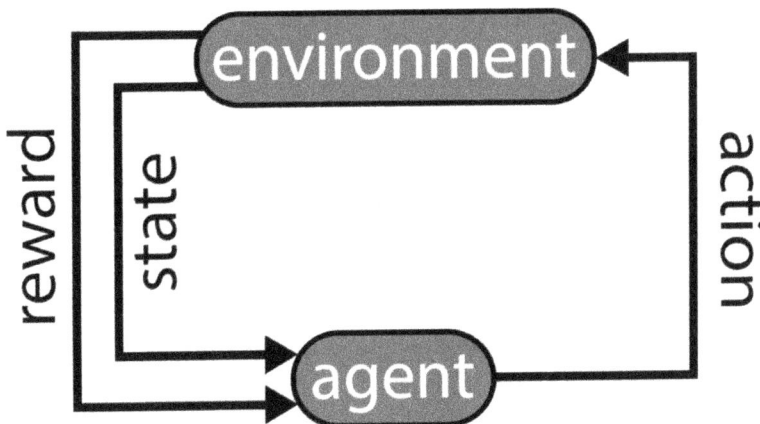

Figure 4.
Basic principle of reinforcement learning (modified after [4]).

analysing the current state of the environment and whether or not its past actions were successful.

Classical applications are board-games (e.g. chess, GO), but there is growing interest in RL for industrial applications (e.g. process optimization).

4. Examples for machine learning tunnel applications

4.1 Automatic rock mass classification approach for TBM excavations

The Brenner Base Tunnel (BBT) which is currently under construction, is a railway tunnel between Austria and Italy, connecting the cities Innsbruck and Fortezza. Including the Innsbruck railway bypass, the entire tunnel system through the Alps is 64 km long and is therefore the longest underground rail link in the world. The BBT consists of a system of two single-track main tunnel tubes, 70 meters apart, that are connected by crosscuts every 333 meters.

A service and drainage gallery lies about 10–12 meters deeper and between the main tunnel tubes (**Figure 5**). During construction the service tunnel serves as an exploratory tunnel, which is driven in advance to gather relevant information about the geology and the expected rock mass behaviour for the main excavation.

The present chapter focuses on 15 km of TBM – operational data from the exploratory tunnel "Ahrental – Pfons", which is part of the construction lot "Tulfes-Pfons". This tunnel section is driven with an open gripper TBM. Throughout the tunnel, the "Innsbrucker Quartzphyllite" and units of the "lower-" and "upper" Schieferhülle" are the dominating lithological units. The rocks consist of low grade metamorphic phyllites to medium grade metamorphic schists with isolated bodies of gneiss, marble and greenschist. During excavation, the rock is mostly of good quality, however, friable and squeezing behaviour as well as large discontinuity driven overbreaks have occurred.

Efforts are undertaken to correlate the data from the exploratory tunnel with the encountered geology with the aim of deriving the rock mass behaviour from the TBM

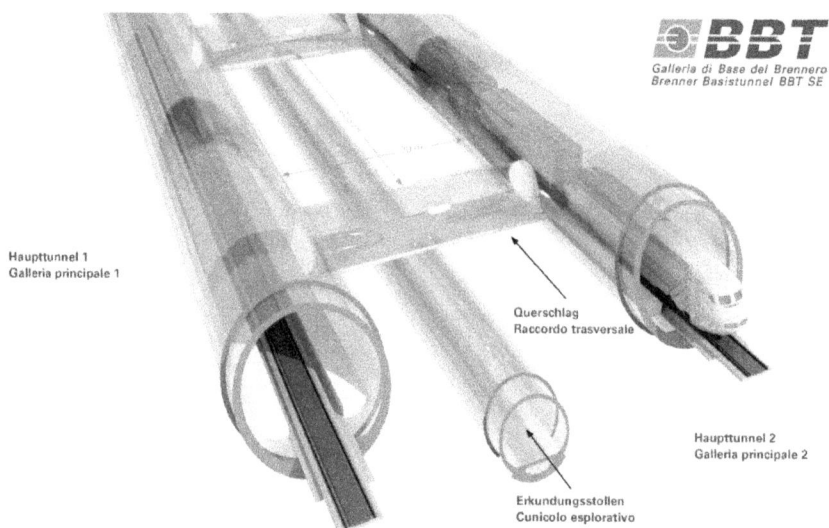

⊜BBT
Galleria di Base del Brennero
Brenner Basistunnel BBT SE

Haupttunnel 1
Galleria principale 1

Querschlag
Raccordo trasversale

Haupttunnel 2
Galleria principale 2

Erkundungsstollen
Cunicolo esplorativo

Figure 5.
Overview of the tunnel arrangement of the BBT [18].

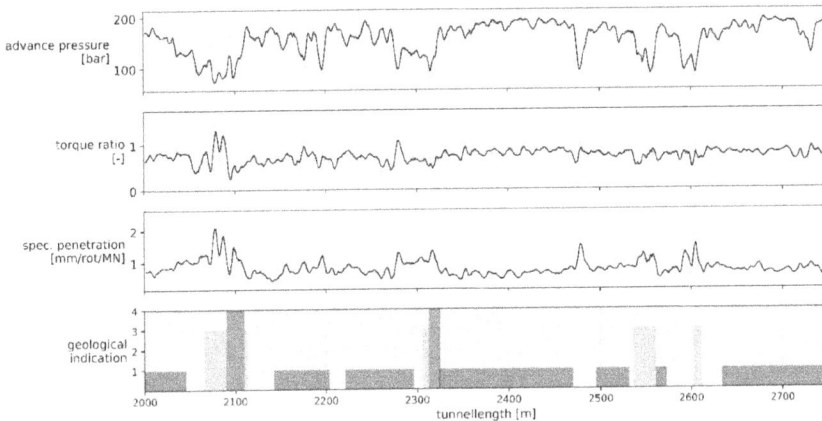

Figure 6.
Exemplary section of TBM data between chainage 2000 and 2750 m; several features show a distinctive response to the encountered fault zone (taken from [2]).

operational data of the main tubes [19]. The TBM data comprises different recorded parameters such as advance force and cutterhead torque or computed parameters like the specific penetration or the torque ratio (after [20]). A corresponding classification of the rock mass behaviour – called the Geological Indication [21] – was also developed and shows the rockmass' quality based on a traffic light system (**Figure 6**). Treating TBM data as input and the rock mass classification as output is a classic application of supervised machine learning. In [1], two different ANNs are given the job to automatically classify TBM operational data into various rock mass behaviour types. In [2], the applicability of a long short term memory networks [22] - a certain type of ANN for sequential data - for the classification of rock mass into behaviour types based on TBM data is shown. In [3], it is shown how an AI system can be misused to get either an optimistic or a pessimistic rockmass classification that might be in favour of one or another party at a specific construction site.

The labels of the geotechnical documentation have been altered to represented a binary form (one-hot encoded vectors), e.g. green = class 1 = [1, 0, 0, 0] (see [1]). Succeeding results show the outcome of applying such a network to the task of automated classification of TBM data (for details see [1]). Between 10,000 and 12,000 tunnel metres of TBM data has been used for training in the above given studies. **Figure 7** shows a result for chainage 1000 to 2000 m. In the upper row, the TBM data (normalised torque ratio) is given, the second row shows the "ground truth" which is the human classification. The third row shows the respective categorical classification of the LSTM network. The resulting output of the final layer (i.e. represented by the probability values for individual classes) is shown in the last row and displays an indication of how "sure" the model is about its assigned classes. This implementation of an LSTM shows adequate accuracies and good consensus between the model and the classification done by humans on site. Where the categorical classification makes the output directly comparable with the human classification more in depth information can be obtained from the probability values resulting from the direct output of ANN.

4.2 Investigation of rock loads via TBM operational data during standstills

Remote rock load monitoring allows TBM operators, engineering geologists and geotechnical engineers to collect, store and process information about the load

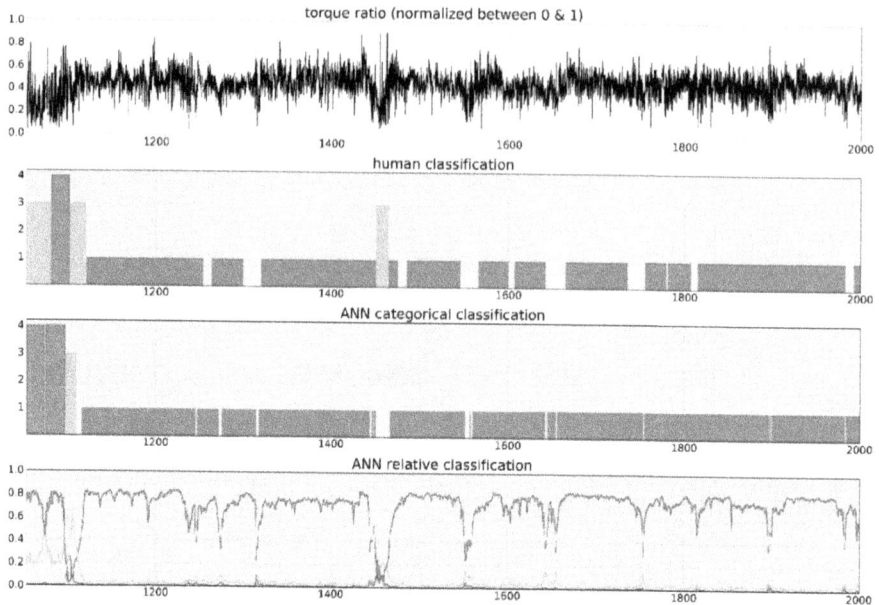

Figure 7.
LSTM network classification from chainage 1000 to 2000 m (taken from [1]).

acting at the interface between TBM shield and the surrounding rock mass, a region that cannot be observed by other expeditious means. It's importance not only lies in the consideration of squeezing ground conditions [23–25], but furthermore in terms of the deformation behaviour and stress redistribution of the surrounding rock masses in hard rock tunnels. To gather relevant information from the collected TBM operational data application of digital systematic data analysis is inevitable.

Many open gripper TBMs are equipped with a roof support shield directly behind the cutterhead which is extended against the tunnel wall during standstills. TBM specifics vary between manufacturers, one example on data logged during the operation of an Herrenknecht open gripper TBM is presented in this chapter. On this machine the roof support shield is driven by two independently movable left- and right cylinders [26]. Sensors separately record the pressure that acts on both sides of the TBM's roof support shield. This provides the unique opportunity to analyse differential rock-loads that are applied to each side of the shield.

Before analysing, the raw data is passed through a pre-processing pipeline with the goal to filter out continuous periods of uninterrupted loading of the shields. Problematically, these loading periods do not simply occur before and after each complete stroke of the TBM, but due to intermediate stops during the excavation process, each stroke is (seemingly) randomly divided into sub-strokes of unequal length. **Figure 8a** gives an example of one stroke, which is separated into five sub-strokes. A blurred analysis would result if the whole stroke was treated as one instead of separating it into sub-strokes.

As throughout the whole tunnel excavation thousands of these sub-strokes would need to be separated, data pre-processing has the goal to achieve a best fitting separation in a fully automated way as manual filtering would be infeasible. A pre-processing pipeline for this problem would consist of the following steps: 1. arranging raw data (e.g. in a database), 2. Filtering out non-advance periods, 3. Checking for and correcting of possible systematic errors, 4. Separating sub-strokes via cluster analysis.

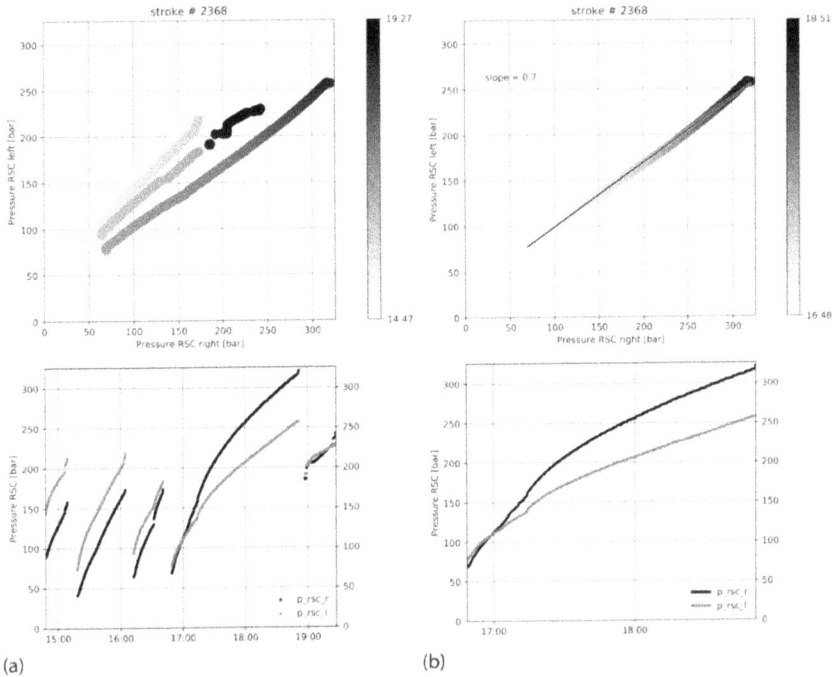

Figure 8.
*Plot of a single complete stroke, in the upper row the pressures in the RSCs left and right have been plotted against each other, whereas in the lower row the pressures were plotted against time ("p_rsc_r" and "p_rsc_l" denotes the pressure in the right and left cylinder respectively). The left column shows (**Figure 8a**) all pressure increases during the stroke and the right column (**Figure 8b**) only shows the longest increase [27].*

After pre-processing, continuous pressure increases for each roof supporting cylinder (RSC) per sub-stroke during standstills of the TBM are isolated (e.g. in **Figure 8b**). In order to do a proper comparison between both RSC's and to take qualitative statements about the stress redistribution/direction in the interface between shield and rock mass, the Line of Isotropic Pressure (LIP) concept [27] is considered.

Plotting the pressures of the left and the right RSC against each other for an isolated sub-stroke (e.g. **Figure 8** upper row), an isotropic pressure increase would represent a straight line of 45°, indicating an equal pressure increase in both cylinders (**Figure 9**). In other words, when fitting a linear regression to the aforementioned plot, the LIP would compare to a regression line with a slope equal to 1. Deviations from the LIP towards the horizontal, corresponding to a decrease in slope equal to values <1, indicate that the pressure increase in the right RSC exceeds the pressure increase in the left cylinder. Same concept applies to deviations from the LIP towards the vertical, corresponding to an increase in slope equal to values >1, indicating that the pressure increase in the left RSC exceeds the pressure increase in the right cylinder. Hence, to assign a slope value to every cluster an extension to the cluster analysis code has to be adapted, fitting a linear least squares regression to every cluster/isolated sub-stroke. At the end of the analysis the data is clustered into significant sub-strokes assigned with a slope value describing the relation of pressure increase between the two RSC's.

Following the approach that the pressure in the RSC's increases with the same extend as the rock load increases, one can state that the rock load acting on the one side of the shield with the higher pressure reading, exceeds the load applied

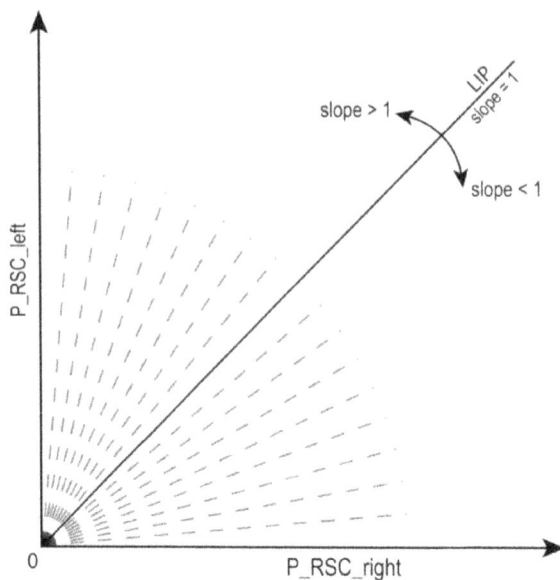

Figure 9.
Conceptual diagram explaining the line of isotropic pressure (LIP): Plot of the pressure in the right RSC on the x-axis vs. the pressure in the left RSC on the y-axis. The LIP corresponds to a linear regression line with a slope of 1 and represents an isotropic increase in pressure in both cylinders [27].

to the shields other side. Plotting the distribution of the slope values in histogram plots either for the whole tunnel, for certain tunnel sections or even parallel to the tunnelling process would hence give a qualitative indication on the rock load distribution in the interface shield to rock mass. In addition to the site characterisation mapped by engineering geologists the pressure in the RSC's provides a vital parameter contributing to the understanding of the overall system behaviour of a tunnel drive.

4.3 Interpretation of monitoring results

Geotechnical monitoring is an integral part of the life cycle of a tunnel structure. The observation method is described in detail in [28]. The observation method is used, on the one hand, to check the design during construction and on the other hand, to check the condition of the tunnel lining during the operational life of the tunnel.

From the technical side, the observational method addresses tunnel surface deformation methods (absolute geodetic measurements, distometers), deformations of the surrounding ground (extensometers) and monitoring of ground support (anchor forces), pressure cells implemented in the shotcrete liner [29].

There are different methods of evaluation and interpretation. The first step is typically the evaluation of a time-displacement diagram. More sophisticated approaches involve the interpretation of displacement vector orientations [29].

Unsupervised ML can be used to develop a warning system for monitoring tunnelling data as it is used today for several other cases of outlier detection (see Section 3.3). This applies to both conventional and machine tunnelling methods. This warning system would consist of a multi-stage pipeline that takes the raw displacement measurements as input and provides an indication of whether a measurement point is behaving 'normally' or not.

4.4 Tunnel maintenance

Many railway and roadway tunnels around the world are ageing. Maintaining works for these tunnels are becoming a major issue. To this day, inspection work is done by visually examining the surface of the lining while walking and through the tunnel and tapping with a hammer on suspicious surfaces (often during night times on temporally closed roads or tracks). Collected data is laborious to process after the inspection.

Digitisation aids this process in terms of making it easier and less subjective. Lately images obtained with different technologies (i.e. laser scanning, slit cameras and line-sensor cameras) find increased usage. These techniques are not only non-destructive, they can also be applied in an automated manner. Especially, vision-based automatic inspection techniques are used to detect damages at the concrete surface of the tunnel lining. In order to recognise and distinguish various types of structural damage of the tunnel lining automatic methods have been introduced [30].

5. Conclusions

Digitisation in general and ML in particular are adding value in tunnelling by improving efficiency of operational processes and quality assurance as well as increasing the safety for on-site personnel by replacing humans with sensors in highly hazardous areas. Nevertheless, these improvements come at the cost of an increasing demand of personnel that is not only skilled in the geotechnical disciplines, but also brings knowledge of ML technology.

The examples given in the previous section show that training ANNs in a supervised manner works and provides valuable information. Nevertheless, today's AI systems – especially the ones based on supervised learning - should only be used as an aid and not as a replacement for geologists or geotechnical engineers on site. The immediate benefit of this technology is the improved classification efficiency and self-consistency but results still need to be critically checked before they are used for decision making. Additionally, ML based automation of the above given processes also increases the safety for human lives and there are also economic advantages that should not be underestimated.

The vision of the "tunneller of the future" who will control the whole construction site and operate all the machines from the comfort of his office chair, with keyboard, joysticks and monitors is still several years ahead of us. To realise this vision, full automation of mechanical underground processes is imperative and to achieve this, great potential is seen in RL technology. The rapid advances in mobile control and navigation technology are giving a sustained boost to automation and robotics in underground mining.

Looking at "evolutionary line for digitalisation in tunnelling" (e.g. [4]), the following developments are foreseeable in the medium term: autonomous machines such as e.g. automatic shotcrete application, autonomous drilling and grouting and driverless dumpers, excavators and loaders for drilling and blasting sequences, real-time adjustments of driving parameters for TBM drives, automatic rock classification procedures, automatic geological updating before the face and e.g. optimised prediction models for sequencing and support quantities. The withdrawal of workers from the most hazardous zones in the active areas of tunnelling is an important aspect of increasing the safety and comfort of underground workers.

Author details

Thomas Marcher*, Georg Erharter and Paul Unterlass
Graz University of Technology, Institute of Rock Mechanics and Tunnelling,
Graz, Austria

*Address all correspondence to: thomas.marcher@tugraz.at

IntechOpen

References

[1] Erharter, G.H., Marcher, T., Reinhold, C., 2019. Comparison of artificial neural networks for TBM data classification, in: Rock Mechanics for Natural Resources and Infrastructure Development- Proceedings of the 14th International Congress on Rock Mechanics and Rock Engineering, ISRM 2019. CRC Press/Balkema, pp. 2426-2433.

[2] Erharter, G.H., Marcher, T., Reinhold, C., 2019. Application of artificial neural networks for Underground construction – Chances and challenges – Insights from the BBT exploratory tunnel Ahrental Pfons. Geomech. und Tunnelbau 12, 472-477. https://doi.org/10.1002/geot.201900027.

[3] Erharter, G.H., Marcher, T., Reinhold, C., 2020. Artificial Neural Network Based Online Rockmass Behavior Classification of TBM Data, in: Springer Series in Geomechanics and Geoengineering. Springer, pp. 178-188. https://doi.org/10.1007/978-3-030-32029-4_16.

[4] Marcher, T., Erharter, G.H., Winkler, M., 2020. Machine Learning in tunnelling – Capabilities and challenges. Geomech. und Tunnelbau 13, 191-198. https://doi.org/10.1002/geot.202000001

[5] Horner, J.; Naranjo, A.; Weil, J. (2016) Digital data acquisition and 3D structural modelling for mining and civil engineering – the La Colosa gold mining project, Colombia in: Geomechanics and Tunnelling 9, pp. 52-57. https://doi.org/10.1002/geot.201500046.

[6] Shin, D.H.; Dunston, P.S. (2008) Identification of application areas for Augmented Reality in industrial construction based on Technology suitability in: Journal of Automation in Construction 17, pp. 882-894.

[7] Zhou, Y.; Luo, H.; Yang, Y. (2017) Implementation of augmented reality for segment displacement inspection during tunneling construction in: Automation in Construction 82, pp. 112-121.

[8] Behzadan, A.H.; Kamat, V.R. (2017) Integrated information modeling and visual simulation of engineering operations using dynamic augmented reality scene graphs in: Journal of Information Technology in Construction, 16, pp. 259-278.

[9] Rankohi, S.; Waugh, L. (2013) Review and analysis of augmented reality literature for construction industry in: Visualization in Engineering (2013).

[10] Goodfellow, I.; Bengio, Y.; Courville, A. (2016) Deep Learning, www.deeplearningbook.org [Accessed: 2021-02-24].

[11] Praveena, M.; Jaiganesh, V. (2017) A Literature Review on Supervised Machine Learning Algorithms and Boosting Process in: International Journal of Computer Applications 169, No. 8, pp. 32-35.

[12] Choi, J. et al. (2012) Combining landslide susceptibility maps obtained from frequency ratio, logistic regression, and artificial neural network models using ASTER images and GIS in: Engineering Geology 124, pp. 12-23.

[13] Javadi, A.A.; Rezania, M. (2009) Intelligent finite element method: An evolutionary approach to constitutive modelling in: Advanced Engineering Informatics 23, No. 4, pp. 442-451.

[14] Konaté, A.A. et al. (2015) Application of dimensionality reduction technique to improve geophysical log data classification performance in crystalline rocks in: Journal of

Petroleum Science and Engineering 133, pp. 633-645.

[15] Bousquet, O., Luxburg, U. and Rätsch, G. (2003) Advanced Lectures on Machine Learning: ML Summer Schools 2003, Canberra, Australia, February 2 – 14, 2003, Tübingen, Germany, August 4 – 16, 2003, Revised Lectures, Lecture Notes in Computer Science, Vol. 3176, Springer, Berlin, Heidelberg.

[16] Zhang, Q.; Liu, Z.; Tan, J. (2019) Prediction of geological conditions for a tunnel boring machine using big operational data in: Automation in Construction, Vol. 100, (2019), pp. 73-83.

[17] Raschka, S., 2017. Python machine learning: Machine learning and deep learning with Python, scikit-learn, and TensorFlow, Community experience distilled, Second edition, fully revised and updated. Packt Publishing, Birmingham, UK.

[18] BBT SE, 2021. "Brenner Base Tunnel", available at: ttps://www.bbt-se.com/en/ (accessed 25 February 2021).

[19] Bergmeister, K.; Reinhold, C. (2017) Learning and optimization from the exploratory tunnel – Brenner Base Tunnel / Lernen und Optimieren vom Erkundungsstollen – Brenner Basistunnel in: Geomechanics and Tunnelling 10, pp. 467-476. https://doi.org/10.1002/geot.201700039.

[20] Radoncic, N.; Hein, M.; Moritz, B. (2014) Determination of the system behaviour based on data analysis of a hard rock shield TBM / Analyse der Maschinenparameter zur Erfassung des Systemverhaltens beim Hartgesteins-Schildvortrieb in: Geomechanics and Tunnelling 7, pp. 565-576. https://doi.org/10.1002/geot.201400052.

[21] Reinhold, C., Schwarz, C., Bergmeister, K., 2017. Development of holistic prognosis models using exploration techniques and seismic prediction. Geomechanik und Tunnelbau 10 (6), 767-778.

[22] Hochreiter, S.; Schmidhuber, J. (1997) Long Short-Term Memory in Neural Computation, No. 9, pp. 1735-1780.

[23] Ramoni, M., Anagnostou, G., 2006. On the feasibility of TBM drives in squeezing ground. Tunn. Undergr. Sp. Technol. 21, 262. https://doi.org/10.1016/j.tust.2005.12.123.

[24] Ramoni, M., Anagnostou, G., 2010. Tunnel boring machines under squeezing conditions. Tunn. Undergr. Sp. Technol. https://doi.org/10.1016/j.tust.2009.10.003.

[25] Ramoni, M., Anagnostou, G., 2011. The interaction between shield, ground and tunnel support in TBM tunnelling through squeezing ground. Rock Mech. Rock Eng. 44, 37-61. https://doi.org/10.1007/s00603-010-0103-8.

[26] Flora, M., Grüllich, S., Töchterle, A., Schierl, H., 2019. Brenner Base Tunnel exploratory tunnel Ahrental-Pfons – interaction between tunnel boring machine and rock mass as well as measures to manage fault zones. Geomech. Tunn. 12, 575-585. https://doi.org/10.1002/geot.201900044.

[27] Unterlass, P., Erharter, G.H., Marcher, T. (2021) Identifying rock loads on TBM shields during standstills (non-advance-periods). Manuscript submitted for publication.

[28] British Standards Institution. (1995). Eurocode 7: Part 1, General rules (together with United Kingdom national application document). London: British Standards Institution.

[29] Schubert, W. et al. (2014) Geotechnical Monitoring in Conventional Tunnelling Handbook.

Austrian Society for Geomechanics,
Salzburg.

[30] Y. Dong et al., "A Deep-Learning-Based Multiple Defect Detection Method for Tunnel Lining Damages," in IEEE Access, vol. 7, pp. 182643-182657, 2019, doi: 10.1109/ACCESS.2019. 2931074.

Chapter 5

Support System Design for Deep Coal Mining by Numerical Modeling and a Case Study

Shankar Vikram, Dheeraj Kumar
and Duvvuri Satya Subrahmanyam

Abstract

Importance of numerical modeling in mine design gained pace after modern way of approach took birth through many variants. Methods such as Continuum and Discontinuum emerge as most effective in resolving certain issues. Cases such as heterogeneity, prevailing boundary conditions in continuum case and presence of discontinuities in other have provided solutions for many causes. A suitable support system is designed for deep virgin coal mining blocks of Godavari Valley Coalfield in India. This analysis is carried out using numerical modeling technique. The results show that the stresses at an angle to the level galleries are adverse. The level gallery/dip-raise may be oriented at 20^0 to 40^0 to reduce roof problems.

Keywords: underground mining, Bord and pillar mining, finite element method, horizontal stress, rock mass classification

1. Introduction

Underground excavation results to stress redistribution and large-scale movement of the roof strata. Therefore, the study on stress is critically important to develop techniques for efficient coal mining [1–6].

In Pench mining area at Thesgora mines where intrusive of basalt flows and faults found, it has been witnessed that high horizontal stress affects the stability of development galleries. After reorientation of dip galleries closer to the principal stress in horizontal direction, no bed dilation was observed in the roof strata of the dip galleries, with improvement in working conditions [7].

This chapter aims to summarize the stress redistribution analyses, which were conducted by the numerical simulation method and design temporary supports based on the horizontal stresses estimated by numerical and empirical methods. The tension-weakening model was adapted for the numerical analysis of rock mass.

2. Details of the work site

The study area, Mandamarri shaft block sector-B is in the northern part of Bellampalli coal belt and it lies in dip side of block. Sullavai formation is the basement rock. The block is covered by barren measure and lower kamthi formation. The trend of the coal seams established from the sub-surface data shows the strike as

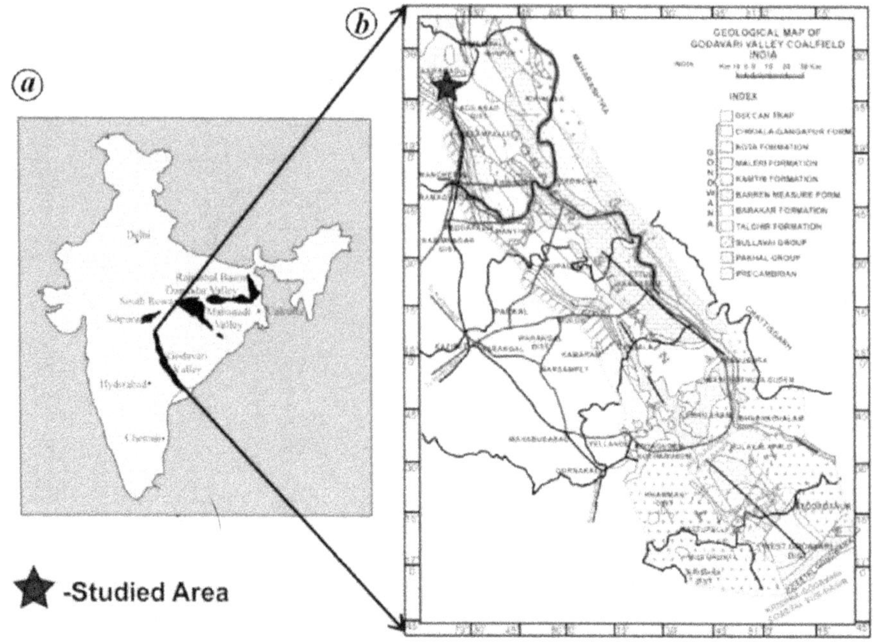

Figure 1.
Location of the investigation area.

North-West to South-East with North-Easterky dipping (**Figure 1**). Coal seam gradient varies from 1 in 3.6 to 1 in 4.3. Three faults have been deciphered sub-surface data.

The Pranahita–Godavari valley coalfield defines a north–northwest–south–southeast trending basin on a Precambrian platform. It is located within the 350 km course of the Pranahita and the Godavari rivers. Bellampalli coal belt comprises of 8 coal seams spread across 38.62 sq.km of 92.54MT.

3. Methodology and calculation sequences

The unfavorable orientation of the mine roadways with respect to high horizontal stress is suspected to be the cause of the roof falls. It is also observed that these roof falls do not occur throughout the mines at the same level though there is no change in the orientations of these roadways. The reason for such observation may be (1) due to favorable orientation of the roadways with respect to the maximum horizontal stress direction, or (2) reorientation of the horizontal stress due to the influence of discontinuities like major faults [8, 9].

Numerical simulation is a powerful technique for studies on rock mechanics and engineering, but its accuracy and reliability lie on the used simulation approach, constitutive model, material properties etc. The finite element method is a numerical solution, divided into non-overlapping regions connected to each other through points called nodes. The behavior of each element satisfying equilibrium conditions, compatibility, material constitutive behavior and boundary conditions is described, and the elements are assembled.

With the numerical simulation method, many studies were conducted on the stress redistribution induced by mining and other factors, among which the inherent perfect elastoplastic and strain-softening models using Mohr–Coulomb failure criterion are most used. However, both constitutive models embedded in FLAC3D (**Table 1**) [10–12].

Case	S_H	Deformation	S_s	S_d
i. (S_H) parallel to level gallery	18.00	12.52	6.00	1.75
ii. (S_H) is 40^0 to level gallery	18.00	12.53	6.50	1.75
iii. (S_H) is perpendicular (85^0)	20.00	12.88	7.50	2.00

Table 1.
Observations at the level gallery /dip-raise.

The safety factor (SF) for supports is estimated by the Eq. (3).

$$P_r \text{ in } t/m^2 = \gamma BF\left[1.7 - 0.037RMR + 0.0002\left(RMR^2\right)\right] \tag{1}$$

$$ASL = nA/Wa \tag{2}$$

$$SF = ASL/P_r \tag{3}$$

4. Model description and simulation

The parameters for boundary conditions were based on *in-situ* stress measurement conducted at study area, and the properties of the rock masses were based on the laboratory tests. To simulate the *In-situ* stress state, a 8.83 MPa load was vertically applied to the top boundary; according to the in situ stress measurement. A horizontal load of 6.22 MPa was applied perpendicular to the direction of strike of coal seam. Along the direction of strike, a horizontal load of 12.44 MPa was

Principal stresses	Results
Vertical Stress (S_v) in MPa (Calculated with an overburden of 517.55 m and density of rock = 2400 kg/m^3	12.17
S_H	12.44 ± 0.16
S_h	6.22 + 0.08
S_H orientation	40^0
K = S_H/ S_v	1.22

Table 2.
Principal stress tensors as evaluated for the study area.

Properties	Coal	Non-Coal
Density (Kg/m^3)	1510	2290
Bulk Modulus K (GPa)	2.12	9.66
Shear Modulus G (GPa)	0.99	4.46
Cohesion C (MPa)	2.0	2.30
The angle of Friction φ (Degree)	20	34
Tensile strength (MPa)	1.0	0.25

Table 3.
Different input parameters considered for the simulation.

considered. The *in-situ* stresses, which were taken into account in the model, are given in **Table 2**. The rock mass properties for the simulation were estimated from the intact rock properties, as summarized in **Table 3**.

5. Analyses and discussion

Study conducted in Australian coal mines has established a relation between roof failure in the roadways and the angle between the roadway axis and the maximum horizontal stress direction. From this a favorable direction of dip and level galleries with respect to major horizontal principal stress direction can be achieved. In Bord and Pillar mining method the dip drives and level galleries are driven perpendicular to each other. In a set of direction of maximum horizontal stress, either one of these or both may be oriented unfavorably with the orientation of the maximum horizontal stress [13–18]. The same has been taken into reference in this study.

A detailed investigation is carried out by numerical modeling to establish the most favorable direction of the dip drives/level galleries vis a vis direction of maximum principal horizontal stress from the stability point of view & design suitable support system.

As a result of numerical analyzing, redistribution of major principal stress (S_H) are given in **Figure 2** for three separate cases. The maximum stress at the roof is observed for case 3 (when Maximum Horizontal Stress is at 85^0 to orientation of level gallery/dip raises). The minimum principal stress at the roof is observed for case 1 (when Maximum Horizontal Stress is parallel to orientation of level gallery/dip raises) (**Table 4**).

The results of numerical analyses for roof convergence are shown in **Figure 3** for three cases. The maximum deformation at the roof is observed for case 3 (when Maximum Horizontal Stress is at 85^0 to orientation of level gallery/dip raises). The minimum deformation at the roof is observed for case 1 (when Maximum Horizontal Stress is parallel to orientation of level gallery/dip raises). The maximum deformation value and its location is introduced in **Table 4** with those of other cases.

In **Figure 4**, the results of numerical analyses on redistribution of shear stresses are given for all cases. The analyses indicate that the case 3 is also critical when considered shear stresses at 85^0 (**Table 4**).

The results of numerical analyzing on shear displacements under loading conditions are shown in **Figure 5**. Maximum shear displacement value and its location is given in **Table 4** with those of other cases.

In the context of this study, numerical simulations have been performed for estimating the major horizontal principal stress, roof displacement, shear stress,

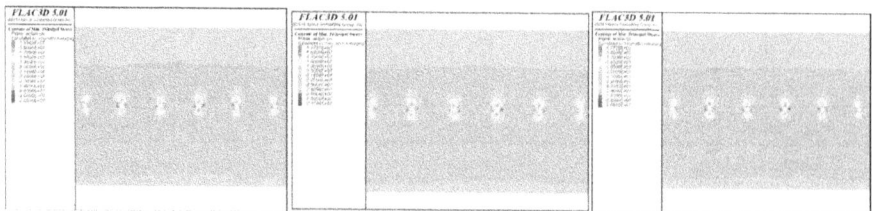

Figure 2.
Distribution of major principle stress: (a) case 1- maximum horizontal stress, which is parallel to orientation of level gallery/dip raises, (b) case 2- max. Horizontal l stress is perpendicular to orientation of level gallery / dip rises, and (c) case 3- max. Horizontal stress, which is 40⁰ to orientation of level gallery/dip rises.

Layer	Rock type	Density (t/cum)	Layer thickness		Structural features			Weatherability%		Strength (MPa)		GW (ml/min)		RMR Value	Classification	
			cm	Rating	Index	Rating		Value	Rating	Value	Rating	Value	Rating		Class	Description
1	MGSST	2.18	30	24	5	20		82	8	0.85	0	10	8	49	III	FAIR
2	MTCGSST	2.2	52	27				91	11	11.9	4			57		
3	FTMGSST	2.23	49	27				88	10	13.3	4			56		
4	CTVCGSST	2.2	69	30				89	10	12.1	4			58		

Table 4.
Weighted RMR evaluated for the strata.

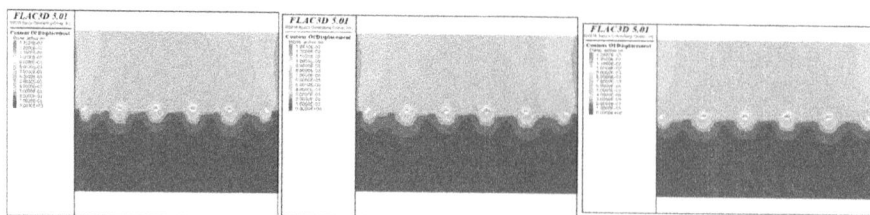

Figure 3.
Distribution of displacement: (a) case 1- maximum horizontal stress, which is parallel to orientation of level gallery/dip raises, (b) case 2- max. Horizontal l stress is perpendicular to orientation of level gallery/dip rises, and (c) case 3- max. Horizontal stress, which is 40° to orientation of level gallery/dip rises.

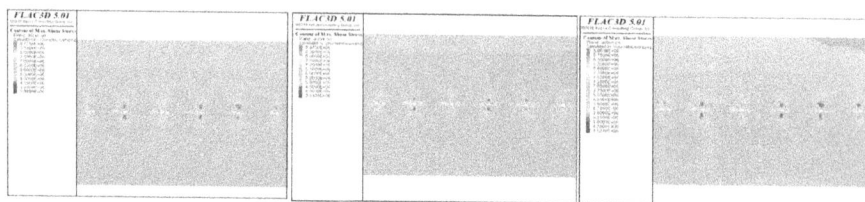

Figure 4.
Distribution of shear stress: (a) case 1- maximum horizontal stress, which is parallel to orientation of level gallery/dip raises, (b) case 2- max. Horizontal l stress is perpendicular to orientation of level gallery/dip rises, and (c) case 3- max. Horizontal stress, which is 40° to orientation of level gallery/dip rises.

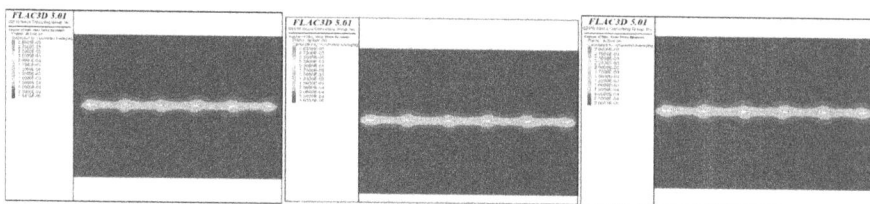

Figure 5.
Distribution of shear displacement: (a) case 1- maximum horizontal stress, which is parallel to orientation of level gallery/dip raises, (b) case 2- max. Horizontal l stress is perpendicular to orientation of level gallery/dip rises, and (c) case 3- max. Horizontal stress, which is 40° to orientation of level gallery/dip rises.

and shear displacement on different mine geometries. The changes for each item have been showed in **Figure 6** on the basis of gallery orientation. The analyses indicate that the level gallery/dip-raise should be oriented at $20°$ to $40°$ to reduce roof problems. As based on the analyses, the authors recommended a temporary support system consisting of bolts for cool mine roof (**Table 5**). The recommend support system is illustrated in **Figure 7**.

6. Conclusion

Support design for an underground opening can only be assessed in conjunction with rock types and structural features. The strength of the rock depends on primarily the in-situ and mining induced stresses. In a common design, analysis begins with evaluation of the strength of the structural features and the forces acting during the mining processes [19].

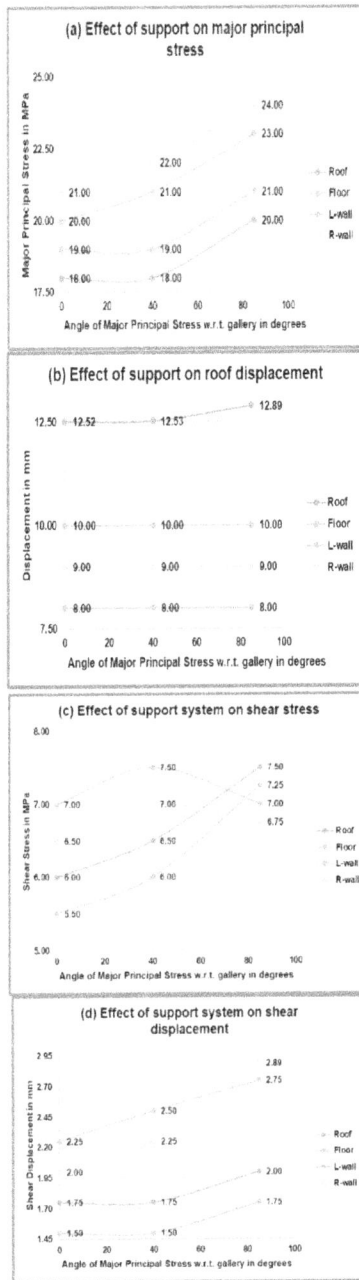

Figure 6.
Changes on the related item as based on orientation: (a) major horizontal principal stress, (b) roof displacement, (c) shear stress and (d) shear displacement.

An underground opening, analysis of the stress distribution is conducted through numerical modeling for different mine geometries. For typical studies, there are certain input parameters, which has to be assessed in field conditions I.e., in-situ measurements with geotechnical studies for the mining blocks. The numerical analyses indicate that the level gallery/dip-raise should be oriented at 20^0 to 40^0 to reduce roof problems.

Recommended Support Details
• Roof Bolts 1.8 M Length 22 mm diameter
• Spacing 1.0 M across and along with galleries
• Bolt density 7750 kg/m3, Young's modulus 2e11 N/m, Tensile strength 1.65e5 N/m.

Table 5.
Support recommendation for coal mine block.

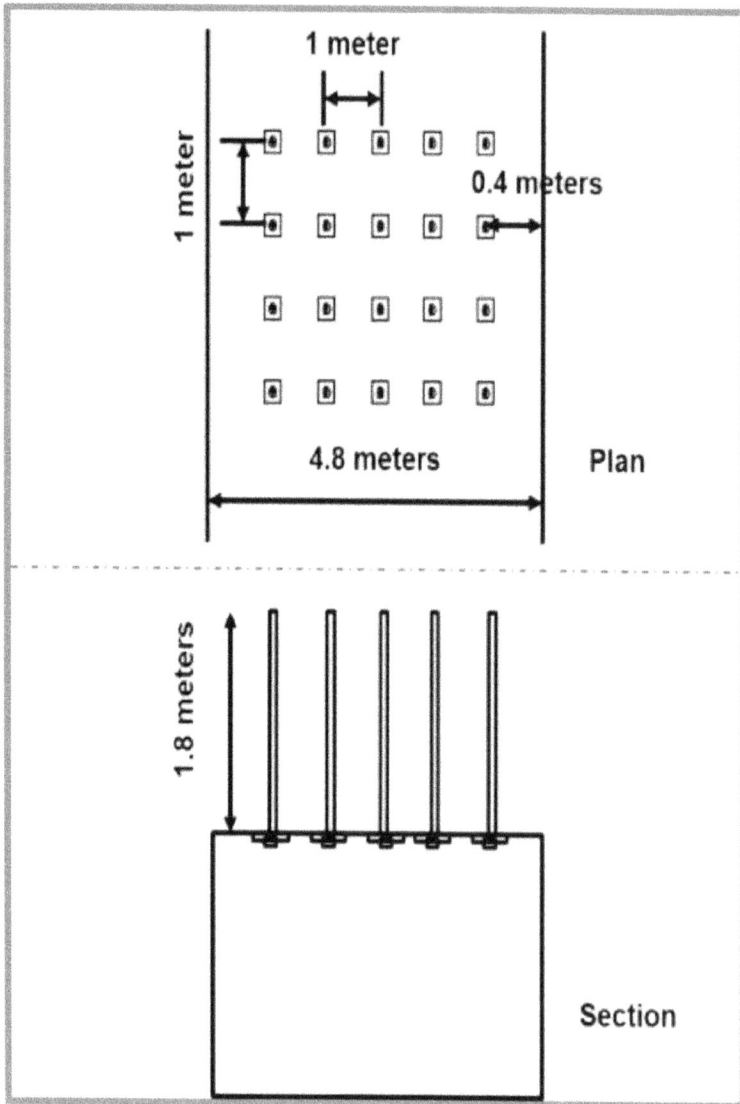

Figure 7.
Support system recommended for roof stability as based on the analyses throughout this study.

Support System Design for Deep Coal Mining by Numerical Modeling and a Case Study
DOI: http://dx.doi.org/10.5772/intechopen.97840

Author details

Shankar Vikram[1*], Dheeraj Kumar[2] and Duvvuri Satya Subrahmanyam[1]

1 National Institute of Rock Mechanics, Bangalore, India

2 Indian Institute of Technology Dhanbad, India

*Address all correspondence to: ajayvaish007@gmail.com

IntechOpen

References

[1] Kushwaha A, Singh SK, Tewari S, Sinha A. Empirical approach for designing of support system in mechanized coal pillar mining, International Journal of Rock Mechanics & Mining Sciences, 47, 2010. p.1063-1078.

[2] Agapito JFT, Gilbride LJ. Horizontal Stresses as Indicators of Roof Stability, SME Annual Meeting Feb.25-27, Phoenix, Arizona, Preprint, 2002. p. 02-056.

[3] Aggson J R and Curran J. Coal Mine Ground Control Problems Associated with a High Horizontal Stress Field, SME Trans. V266. 1979. pp 1972-1978.

[4] Amedi Bernard and Stephanson Ove. Rock Stress and its Measurement, Chapman and Hall Publishers, London, 1997.

[5] Enever, J. R.; Walton, R. J.; Windsor, C R. Stress regime in the Sydney Basin and its implications for excavation design and construction. In: CSIRO Division of Geomechanics, Mt Waverley, editor/s. The Underground Domain: Seventh Australian Tunnelling Conference; Sep 11-13, 1990; Sydney, N.S.W. Sydney, N.S.W.: Institution of Engineers Australia; 1990. 49-59.

[6] Fairhurst C. In-Situ Stress Determination an Appraisal of its Significance in Rock Mechanics, Proc. Intl. Symp. On Rock Stress and Rock Stress Measurements, Stockholm, Centek Publ., Luela, 1986. pp 3-17.

[7] Gale WJ, Fabjanczyk MW. Strata Control Utilizing Rock Reinforcement Techniques in Australian Coal Mines, Symposium on Roof Bolting, Poland, 1991.

[8] Gilbride L J, Agapito J F T, Kehrman R (2004): Ground Support Design Using Three-Dimensional Numerical Modelling at Molycorp, Inc's, Block Caving Questa Mine, Mass. Min Chile 2004, Santiago Chile.

[9] Hart R. Enhancing Rock Stress Understanding through Numerical Analysis, International Journal of Rock Mechanics and Mining Sciences, 40, 2003. pp 1089-1097.

[10] Hoek E, Brown ET. Underground Excavations in Rock, Institution of Mining and Metallurgy, 1980. p. 100, 183-241.

[11] Kong P, Jiang L, Jiang J, Wu Y, Chen L, Ning J, Numerical Analysis of Roadway Rock-Burst Hazard under Superposed Dynamic and Static Loads. Energies, 12, 2019. p. 3761.

[12] Manohara Rao, Sharma DN. Stress Orientation in the Godavari Gondwana Graben, India, Journal of Rock Mechanics & Tunnelling Technology, 20(2), 2014. p. 109-119.

[13] Qian M, Xu J. Study on the "O shape" circle distribution characteristics of mining induced fractures in the overlying strata. Journal of China Coal Society, 23(5),1998. p. 466-9.

[14] Rocscience. 3D Meshing Customization Developers Tips. 8 Rocscience Inc. 2013.

[15] Rocscience. Personal Communication, 2014a.

[16] Rocscience. Phase2 Theory-Convergence Criteria. 4. 2014b.

[17] Sanyal K, Bahadur AN. A Geotechnical Study for Roof Control of Thesgora Mine, Pench Area, Western Coal Fields Ltd., National Conference on Ground Control in Mining, BHU. 1996.

[18] Shengwei Li, Mingzhong Gao, Xiaojun Yang, Ru Zhang, Li Ren, Zhaopeng Zhang, Guo Li, Zetian Zhang, Jing Xie. Numerical simulation of spatial distributions of mining-induced stress and fracture fields for three coal mining

layouts. Journal of Rock Mechanics and Geotechnical Engineering. 10: (2018). p. 907-913. https://doi.org/10.1016/j.jrmge.2018.02.008.

[19] Swoboda G and Marence M (1992) : Numerical Modelling of Rock Bolts in Intersection with Fault System, Numerical Models in Geomechanics, Pande and Petruszczak (ecs) © 1992 Balkema, Roterdam.

www.ingramcontent.com/pod-product-compliance
Lightning Source LLC
Chambersburg PA
CBHW081241190326
41458CB00016B/5867